D1284588

*Paul W. Unger, PhD*

# Soil and Water Conservation Handbook
## *Policies, Practices, Conditions, and Terms*

WITHDRAWN

*Pre-publication*
*REVIEWS,*
*COMMENTARIES,*
*EVALUATIONS . . .*

"Dr. Paul Unger, world-known for his research on soil and water conservation, has written a valuable and unique book—a dictionary of terms used in soil and water conservation. He has based his definitions on forty years of working in the area, plus the knowledge he gained from his father and grandfather, who farmed with the challenges of erosion. This is a significant contribution to the literature that I am going to place on the top of my desk right beside my *Webster's Dictionary.* I recommend this book to everyone involved in agriculture—policymakers, teachers, researchers, engineers. It is a treasure of the author's lifetime's knowledge shared with us. It has opened my eyes to a wide view of agriculture and the many definitions that are possible. This book will become a classic."

**M. B. Kirkham, PhD**
*Professor, Department of Agronomy, Kansas State University, Manhattan*

"This handbook is a state-of-the-knowledge compendium of basic principles and terminology pertinent to prominent environmental issues of soil erosion by water and wind, surface runoff, dissolved and suspended loads, and nonpoint source pollution. It is also an excellent reference source for soil and water conservation practices. The handbook is comprehensive, all-encompassing, up-to-date, and reader friendly. This valuable reference source will be of professional interest to students, practitioners, academicians, policymakers, and the general public interested in the stewardship of soil, water, air, and other natural resources. The technical material presented is authentic, credible, and of high scientific quality. This is a valuable book for anyone interested in the science, practice, and policy of soil and water conservation and management."

**Rattan Lal**
*President-Elect, Soil Science Society of America; Professor, Director, Carbon Management and Sequestration Center, The Ohio State University*

*More pre-publication*
*REVIEWS, COMMENTARIES, EVALUATIONS . . .*

"This handbook has a fortunate combination of the author's clear-headed, analytical approach, sense for integration, and direct access to and contact with a wide range of different agroecological and socioeconomic contexts. It is addressed to readers who want to know how to make our agricultural practices a sustainable and safe set of activities as the demands increase. It also provides information that balances the bias toward biochemistry and genetics by emphasizing land, soil nutrients, and water as depletable resources for mankind and by emphasizing the ecosystem services to society: the protection of watershed, the recycling of water, and the maintenance of biodiversity."

**José R. Benites, PhD**
*FAO Representative in Argentina*

"This work by Dr. Unger is a very comprehensive compendium of information that covers a broad spectrum of subjects related to soil and water conservation. It sets a benchmark that will help standardize terminology in a way that will truly improve communication between the farm community, the general public, and policymakers who may not have a background in agriculture. The alphabetical arrangement of the topics in the handbook makes it easy to use, especially for the nonagriculturist. This handbook represents a lifelong effort by one of the 'greats' in the discipline of soil and water conservation. There is no other work like it to the best of my knowledge, and it is a legacy we will all appreciate into the long-term future."

**Gary A. Peterson, PhD**
*Professor and Head,*
*Department of Soil and Crop Sciences,*
*Colorado State University*

Haworth Food & Agricultural Products Press™
An Imprint of The Haworth Press, Inc.
New York • London • Oxford

### *NOTES FOR PROFESSIONAL LIBRARIANS AND LIBRARY USERS*

This is an original book title published by Haworth Food & Agricultural Products Press™, an imprint of The Haworth Press, Inc. Unless otherwise noted in specific chapters with attribution, materials in this book have not been previously published elsewhere in any format or language.

### *CONSERVATION AND PRESERVATION NOTES*

All books published by The Haworth Press, Inc., and its imprints are printed on certified pH neutral, acid-free book grade paper. This paper meets the minimum requirements of American National Standard for Information Sciences-Permanence of Paper for Printed Material, ANSI Z39.48-1984.

### *DIGITAL OBJECT IDENTIFIER (DOI) LINKING*

The Haworth Press is participating in reference linking for elements of our original books. (For more information on reference linking initiatives, please consult the CrossRef Web site at www.crossref.org.) When citing an element of this book such as a chapter, include the element's Digital Object Identifier (DOI) as the last item of the reference. A Digital Object Identifier is a persistent, authoritative, and unique identifier that a publisher assigns to each element of a book. Because of its persistence, DOIs will enable The Haworth Press and other publishers to link to the element referenced, and the link will not break over time. This will be a great resource in scholarly research.

St. Louis Community College
at Meramec
LIBRARY

# Soil and Water Conservation Handbook

## *Policies, Practices, Conditions, and Terms*

*HAWORTH FOOD & AGRICULTURAL PRODUCTS PRESS™*
Sustainable Food, Fiber, and Forestry Systems
Raymond P. Poincelot, PhD
Senior Editor

*Biodiversity and Pest Management in Agroecosystems* by Miguel A. Altieri

*Developing and Extending Sustainable Agriculture: A New Social Contract* edited by Charles A. Francis, Raymond P. Poincelot, and George W. Bird

*Soil and Water Conservation Handbook: Policies, Practices, Conditions, and Terms* by Paul W. Unger

# Soil and Water Conservation Handbook
## *Policies, Practices, Conditions, and Terms*

Paul W. Unger, PhD

Haworth Food & Agricultural Products Press™
An Imprint of The Haworth Press, Inc.
New York • London • Oxford

For more information on this book or to order, visit
http://www.haworthpress.com/store/product.asp?sku=5678

or call 1-800-HAWORTH (800-429-6784) in the United States and Canada
or (607) 722-5857 outside the United States and Canada

or contact orders@HaworthPress.com

Published by

Haworth Food & Agricultural Products Press™, an imprint of The Haworth Press, Inc., 10 Alice Street, Binghamton, NY 13904-1580.

© 2006 by The Haworth Press, Inc. All rights reserved. No part of this work may be reproduced or utilized in any form or by any means, electronic or mechanical, including photocopying, microfilm, and recording, or by any information storage and retrieval system, without permission in writing from the publisher. Printed in the United States of America. Reprint - 2007

PUBLISHER'S NOTE
The development, preparation, and publication of this work has been undertaken with great care. However, the Publisher, employees, editors, and agents of The Haworth Press are not responsible for any errors contained herein or for consequences that may ensue from use of materials or information contained in this work. The Haworth Press is committed to the dissemination of ideas and information according to the highest standards of intellectual freedom and the free exchange of ideas. Statements made and opinions expressed in this publication do not necessarily reflect the views of the Publisher, Directors, management, or staff of The Haworth Press, Inc., or an endorsement by them.

Cover design by Lora Wiggins.

**Library of Congress Cataloging-in-Publication Data**

Unger, Paul W.
Soil and water conservation handbook : policies, practices, conditions, and terms / Paul W. Unger.
p. cm.
Includes bibliographical references and index.
ISBN-13: 978-1-56022-329-0 (hard : alk. paper)
ISBN-10: 1-56022-329-4 (hard : alk. paper)
ISBN-13: 978-1-56022-330-6 (soft : alk. paper)
ISBN-10: 1-56022-330-8 (soft : alk. paper)
1. Soil conservation—United States—Handbooks, manuals, etc. 2. Water conservation—United States—Handbooks, manuals, etc. I. Title.
S624.A1.U54 2006
631.4'50973—dc22

2005033802

# CONTENTS

## ABOUT THE AUTHOR

**Paul W. Unger, PhD,** retired in 2000 after 35 years with the United States Department of Agriculture-Agricultural Research Service (USDA-ARS). Most of his research pertained to the effects of tillage systems, cropping systems, and crop residue management practices on soil and water conservation, crop production, and soil conditions under dryland and limited irrigation conditions in the Great Plains region. Dr. Unger has more than 240 publications to his credit, including articles in technical journals, in conference proceedings, and as book chapters. He is the author of *Tillage Systems for Soil and Water Conservation* and editor of *Managing Agricultural Residues*.

© 2006 by The Haworth Press, Inc. All rights reserved.
doi:10.1300/5678_a

# Foreword

Soil, water, and air are the natural resources that sustain life. Soil provides a medium for plant growth, regulates how water moves over and through land areas, and serves as a buffer to environmental change. Soils and the biological, chemical, and physical processes that form them from rocks and other parent materials are vital for the production of food and fiber that are required to sustain humankind.

Soil erosion occurs in every part of the world and threatens crop production and the natural environment. Although soil erosion is a natural process, it is often greatly accelerated by the actions of people. The conversion of natural ecosystems to agricultural ecosystems involves clearing the land of its natural vegetation using intensive tillage or other means that often greatly disturb the soil. In most cases, short-term benefits accrue because these new lands are usually highly productive and yield bountiful harvests. However, the long-term consequences are detrimental in many instances because soil degradation gradually but surely occurs. Maintaining soil sustainability is an ongoing challenge not only for individual farmers but for countries and the world community. According to Franklin D. Roosevelt, the 32nd president of the United States of America, "The history of every nation is eventually written in the way in which it cares for its soil."

In the past few decades, there has been increased interest in reducing tillage and maintaining more crop residues on the soil surface. The benefits are many. The presence of crop residues reduces raindrop impact on the soil surface, increases water infiltration into the soil, reduces soil water evaporation, increases carbon sequestration, improves biological activities, reduces soil compaction, and aids in other processes. In many situations, however, crop residues are used

© 2006 by The Haworth Press, Inc. All rights reserved.
doi:10.1300/5678_b

for feed, fuel, or other purposes and are removed to the fullest extent feasible. This can lead to accelerated soil erosion and degradation.

Dr. Paul Unger is an internationally acclaimed soil scientist who has spent many years studying, lecturing, and writing about the effects of tillage, crop residues, and soil amendments on soil and water conservation. He has traveled to more than 40 countries to observe soil and water conservation practices and to exchange ideas and information with other scientists. In addition, he has assembled a large personal library of publications dealing with soil and water conservation. In this book, Dr. Unger has compiled and explained a comprehensive list of policies, practices, conditions, and terms related to soil and water conservation. Terms such as "no-tillage," "reduced tillage," "minimum tillage," "conservation tillage," "limited tillage," and "conservation agriculture" can be confusing and sometimes misleading. This compendium will be useful to researchers, extension agents, policymakers, and others dealing with or interested in soil and water conservation by helping them to understand the complexity and confusion sometimes expressed in the literature.

*B. A. Stewart*
*Director, Dryland Agriculture Institute*
*West Texas A&M University*
*Canyon, Texas*

# Preface

Soil, the thin layer of mineral and organic materials at the earth's surface not covered by water, is the natural medium for growing the plants that provide much of the food, fiber, and other materials used by humans and also provide the main source of food for many animals, birds, and other creatures. Plants serve as shelter for animals and provide for an aesthetic environment. Unfortunately, soil has been and continues to be seriously eroded by water or wind at many sites throughout the world. Erosion results in streambank collapse; sediment deposition at downslope sites, in streams, and in reservoirs; damage to roads, buildings, and other structures; damage to equipment; accidents; and health problems. When erosion occurs, society as a whole is affected because correcting the damage usually involves monetary expenditures, often borne by some governmental agency. Therefore, society as a whole should be concerned when erosion occurs and should support soil conservation efforts.

Closely associated with the need for soil conservation is the need for water conservation, because soil and water losses often occur simultaneously, with water being the soil transport mechanism where erosion by water occurs. When water is limited, as in drier climatic regions and sometimes even in humid regions, the poor plant growth that results often increases the potential for erosion by water and wind. In addition, water conservation is important for agricultural, residential, industrial, and recreational users because competition for water has become a major issue. The competing users, including agriculture, obtain some water from aquifers, but much is derived from precipitation stored in surface reservoirs, and some water for agricultural use is stored in the soil.

© 2006 by The Haworth Press, Inc. All rights reserved.
doi:10.1300/5678_c

Soil and water conservation has received, and continues to receive, much attention, as evidenced by numerous reports in the literature. This book is not intended to be a historical review, but it includes entries for historical and current policies, practices, conditions, and terms related to soil and water conservation. It is essentially a "mini-encyclopedia," with more than 700 main and lower-order entries. The entries are the result of my interest in soil and water conservation, which began early in life, remained with me throughout my research career as a soil scientist with the Agricultural Research Service of the U.S. Department of Agriculture (from which I retired in 2000), and continues to the present. The purpose is to serve as a "ready reference" and source of information for anyone interested in or dealing with any aspect of soil or water conservation. By avoiding discussion that is highly technical, the book is suitable for beginners as well as for experts—that is, for anyone concerned with conserving these natural resources for sustained use for present and future inhabitants of the earth.

# Introduction

Soil, the thin layer of unconsolidated mineral and organic materials at the earth's surface not covered by water, has been and continues to be seriously degraded by erosion by water and/or wind at many sites throughout the world. Soil is the natural medium for plant growth, and humans obtain much of their food, fiber, and various other materials from plants. Plants also are the main source of food (e.g., forages, grains, seeds, tubers, and roots) for animals, birds, and other creatures; they serve as shelter for animals and provide for an aesthetic environment. Conservation of soil, therefore, is highly important.

Soil erosion by water and wind affects many things other than plant growth. The effects include streambank collapse; sediment deposition at downslope sites and in streams and reservoirs; damage to roads, highways, buildings, and other structures; damage to equipment; accidents; and health problems as a result of windborne dust. Erosion affects society as a whole because correcting the resulting damage usually involves monetary expenditures by some government agency. Therefore, society as a whole should be concerned with efforts to control erosion.

Closely associated is the need for water conservation under many conditions because soil and water losses often occur simultaneously. When water is limited, as in the drier climatic regions and sometimes even in humid regions, poor plant growth often leads to conditions that increase the potential for erosion by water or wind. Water conservation is important also for residential, industrial, and recreational users, and competition for water has become a major issue among such users in some regions. These users obtain some water from

© 2006 by The Haworth Press, Inc. All rights reserved.
doi:10.1300/5678_01

aquifers, but much is from precipitation stored in surface reservoirs for later use.

Soil and water conservation has received much attention for many years, and it continues to receive much attention, as evidenced by the numerous reports in the literature. This book is not intended as a historical review of soil and water conservation; it includes entries for historical and current policies, practices, conditions, and terms related to soil and water conservation.

Although I did not recognize it as such at the time, my earliest recollection of erosion is that of windborne soil (dust) in south central Texas in the United States during the drought years of the 1930s. That part of Texas was not affected by the infamous Dust Bowl, but rainfall was low some years, and dust from a nearby field was deposited in our house and farmstead.

As a youngster, I also saw my grandfather trying to control gully erosion where water on his farm flowed from relatively flat upland areas downslope along paths traversed by cattle and farm machinery, including wagons and cultivators. His efforts were largely in vain because he tried to block water flow in the gullies rather than slowing or preventing water flow into the gullies from the upland areas. Some years later, personnel from the United States Department of Agriculture Soil Conservation Service (USDA SCS, now the Natural Resources Conservation Service [NRCS]) designed diversion terraces for his farm, the construction of which controlled the erosion. My father also achieved control of erosion by water on his farm with the assistance of USDA SCS personnel. Through their efforts, diversion terraces, graded field terraces, and appropriate waterways were installed to nonerosively carry excess water from his farm.

My first formal introduction to soil conservation occurred in my Vocational Agriculture class as a freshman in high school, where I learned to use a field level, determine the slope of land, and calculate the appropriate vertical interval (spacing) between adjacent terraces. The exact formula used eludes me, but it was similar to (or may have been)

$$\mathrm{VI} = 2 + S/4,$$

where VI is the vertical interval and $S$ is the slope.

My interest in controlling erosion stems from the erosion on my father's and grandfather's farms and the Vocational Agriculture course. That interest has remained with me over the years, including during my career as a soil scientist with the USDA Agricultural Research Service (ARS) from 1965 until 2000. Much of my research with the ARS dealt with water conservation (water storage in soil for later use by crops) and not directly with soil conservation (erosion control). Conserving water in many cases, however, reduces the amount and rate of runoff, thereby reducing the potential for erosion by water. The conserved water also often improves subsequent plant growth, thereby providing conditions that reduce the potential for erosion by water and wind.

Many factors affect soil and water losses, and numerous reports in the agricultural literature deal with such losses, both from a historical viewpoint and with regard to current information. These factors include the attitudes and capabilities of the person working the land, the policies of national governments, the natural conditions of the land, the climate, and the practices applied to the land. Thus, soil and water conservation is also affected by many factors.

Over the years, I have learned of and collected information about numerous policies, practices, conditions, and terms related to soil and water conservation, which form the basis of this book. Some entries cover topics that are now obsolete but are included because they remain in the literature for historical reasons. Most main and secondary entries are briefly described, together with their effect on soil and/or water conservation. Where for some secondary and lower-level entries the effect on soil and/or water conservation is not given, it is essentially the same as for the main entry. Some entries are cross-referenced to others that have the same meaning or essentially the same effects. Sources for the entries include technical journals, bulletins, and reports; farm magazines; commercial leaflets and bulletins; Internet sites; and the key publications listed as "Further Reading" below. Additional information is available in the literature and on the Internet.

Numerous entries pertain to soil and/or water conservation practices. My intent is to provide basic information regarding the practices but not to recommend any one for application to a particular condition or site. Some practices are suitable or adaptable to a wide range of conditions; others are suitable for specific conditions. Therefore, if one is to achieve the greatest soil and/or water conservation benefits, the conditions at a given site must be carefully evaluated and then the most suitable practice must be applied.

The main entries are given in alphabetical order, with subentries ordered alphabetically below them. Scientific names for plants, animals, and other organisms mentioned are listed in the appendix.

## *Further Reading*

Bennett, Hugh Hammond (1939). *Soil Conservation.* New York and London: McGraw-Hill Book Company, Inc.

Hudson, Norman (1981). *Soil Conservation*, Second Edition. Ithaca, NY: Cornell University Press.

Soil Conservation Society of America (1982). *Resource Conservation Glossary*, Third College Edition. Ankeny, IA: Soil Conservation Society of America.

Soil Science Society of America (1997). *Glossary of Soil Science Terms, 1996.* Madison, WI: Soil Science Society of America.

# ALPHABETICAL LIST OF ENTRIES

# AGRICULTURE

Agriculture is the science and art of farming, which includes the work of cultivating the soil, producing crops, and raising livestock. This definition covers many of the items discussed in this book. Additional comments on conservation, dryland, irrigated, precision, and rainfed agriculture are appropriate here because these types of agriculture strongly influence soil and water conservation.

## *Conservation Agriculture*

Conservation agriculture is a system based on the integrated management of soil, water, and agricultural resources. Its main objective is economical, ecological, and socially sustainable agricultural production while the soil is being regenerated or soil degradation is being reversed. The system is based on maintaining a permanent soil cover, practicing minimal soil disturbance, and using crop rotations. Benefits include fuel savings; implement maintenance and replacement savings; greater and more stable crop yields; labor (time) savings; a lower requirement for heavy work; crop diversification; food security and diet improvements; improved water availability, amount, and quality; improved soil fertility and regeneration; less erosion; improved air quality; and greater agricultural biodiversity (more diverse crop rotations, enhanced soil biodiversity, and less pressure on marginal lands, forests, and natural reserves).

© 2006 by The Haworth Press, Inc. All rights reserved.
doi:10.1300/5678_02

### *Dryland Agriculture*

Dryland agriculture, also known as dryland farming, is crop (and sometimes also animal) production under conditions where limited or erratic precipitation generally results in moderate or severe plant water stress during a substantial portion of the year. Special cultural practices, adapted crops, appropriate cropping systems, and suitable practices to conserve water are required for successful and stable crop production under dryland conditions. Year-round cropping on the same land usually is not possible, but one crop per year often is possible with appropriate management. Limited plant growth means that conditions sometimes are conducive to erosion by wind and that erosion by water may occur during the high-intensity rainstorms that sometimes occur where dryland agriculture is practiced.

### *Irrigated Agriculture*

Irrigated agriculture is widely used in water-deficient (low-precipitation) regions where crop production would otherwise be limited, if not impossible. The water is derived from streams, reservoirs, or aquifers. With proper management, soil and water conservation can generally be achieved. With poor management, soil and water conservation may be poor. Competition for water among different users has resulted in water supplies for irrigation becoming limited in some regions. In addition, the decline in supplies in some aquifers is curtailing irrigated agriculture in some regions (*see* **IRRIGATION**).

### *Precision Agriculture*

Precision agriculture, also known as site-specific management, involves the differential application of inputs in a crop production system, based on spatial and temporal variations that affect crop production within a management unit (field). It is a new concept that can be defined as a comprehensive system designed to optimize agricultural production. Objectives include increased production efficiency, improved product quality, more efficient chemical use, energy conservation, soil and groundwater protection, and soil and water conservation. Precision agriculture begins with crop planning and includes tillage, planting, chemical application, and harvesting operations and postharvest processing of the crop. The system involves basing applications to the soil or crop on predefined maps of soil or crop conditions or on sensor readings generated as the machinery crosses the field. Variables include the soil's water-holding capacity, organic matter content, fertility level, topography, and microbial populations; weed and insect populations; disease occurrence; crop growth; and harvestable yield. Through appropriate applications, improved crop growth and production should be achieved, which usually is beneficial for providing crop cover on the land to reduce the potential for erosion by water and/or wind and to increase the potential for water conservation. Improved production usually also increases the income for the producer or landowner, thus providing financial resources for implementing and maintaining soil and water conservation practices when needed.

## *Rainfed Agriculture*

Rainfed agriculture is crop and animal production for which precipitation (rain or snow) is the sole source of water (no irrigation water is applied). Rainfed agriculture is practiced from humid regions where precipitation usually is frequent, water is plentiful, and water conservation often receives little attention to arid and semiarid regions where precipitation is limited and erratic and water conservation is highly important for successful crop and animal production (*see* **AGRICULTURE, *Dryland Agriculture***). Production without irrigation in arid regions usually depends on capturing the runoff from a relatively large area during infrequent precipitation events and concentrating the water on a relatively small cropped area. The potential for erosion by water exists in all rainfed regions, but the potential for erosion by wind usually is greatest in the dry regions. On some soils and under some conditions, erosion by wind may occur in humid regions.

# ANIMALS

Domestic, wild game, exotic, and other land-inhabiting animals frequently have a major impact on the *potential* for soil and water losses and, consequently, for soil and water conservation. The factors include the nature of their habitat, populations, and feeding habits.

## *Burrowing Animals*

Burrowing animals frequently consume most vegetation surrounding their burrows. Others destroy plant root systems or underground plant parts, thereby destroying the plants. As a result, the denuded soil may be susceptible to erosion by water because of greater runoff and by wind because the surface protection usually provided by plants has been removed. The greater runoff decreases the potential for water conservation. Burrows on sloping land may become the starting point of gullies as a result of water flowing into them during large precipitation events.

## *Grazing Animals*

Grazing animals affect the potential for erosion on a given tract of land according to the closeness to the surface at which they graze the vegetation and by trampling the surface, which may destroy the existing vegetation and retard or prevent its regrowth. Closeness of grazing, to a large extent, depends on the animal. For example, sheep and goats generally graze closer to the surface than cattle.

## *Grazing Control*

Ranges, grasslands, and pastures and fields (under some conditions after harvesting of the crop for its primary product) are grazed by animals. To ensure maintenance of surface conditions conducive to conserving soil and water, grazing must be controlled so that adequate plant cover remains on the soil surface. The advice for many years has been to "take half and leave half" of the vegetation available for grazing. In addition, uniform distribution of grazing on a given tract of land is important to avoid overgrazing on one part (e.g., adjacent to the water supply). Grazing distribution can be improved by developing several watering sites; placing minerals (e.g., salt) at several locations on the tract; and subdividing the tract into smaller areas, then practicing rotation grazing (allowing grazing on an area for a relatively short time before allowing grazing on the next area). A major consideration is to properly match the number of animals to the amount of vegetation available for grazing while maintaining adequate plant materials on the surface for soil and water conservation purposes.

### *Livestock/Forage Supply Ratios*

*See* **ANIMALS, *Grazing Animals; Grazing Control.***

### *Overgrazing*

Overgrazing occurs when animal populations are too large for the amount of forage materials available on the area being used for grazing (rangeland, grassland, and pastures). It reduces the vegetation to such a level that the land may no longer be adequately protected against erosion and the potential for water conservation may be greatly reduced. Short-term overgrazing can be overcome by removing the animals from the given tract of land or by supplying them with forage (hay, silage, etc.) from an outside source. Continued overgrazing commonly leads to land degradation, which often results in severe erosion and poor water conservation.

### *Populations*

When populations of animals, whether wild or domestic, are in line with the carrying capacity of the land being grazed, erosion by water or wind often is not a problem if adequate vegetation is present to provide protection against erosion (*see* **RANGELANDS/GRASSLANDS/PASTURES, *Carrying Capacity***). When animal populations exceed the carrying capacity, the result is overgrazing, which is the main cause of severe erosion by water and wind on noncroplands (native rangelands) in many places. Continued overgrazing weakens the plants, setting in motion a progressive decline in the carrying capacity of the given tract of land. Allowing animals to graze forages and other plant materials on croplands is a common practice, and as on noncroplands, overgrazing may lead to severe erosion by water and wind. In addition, under many conditions, especially when the soil is wet, trampling by grazing animals causes soil compaction, which subsequently may lead to greater runoff and erosion by water because infiltration is decreased. Such conditions also result in decreased water conservation, unless the compacted soil is loosened (e.g., by tillage).

# B

## BARRIERS

Barriers for controlling erosion by water or wind and for improving water conservation can be vegetative (e.g., crop plants, grasses, bushes, and trees) or nonvegetative (e.g., snow fences and terraces). They are used to reduce the flow rate or concentration of water as it flows downslope and to reduce wind speed at the soil surface downwind from the barrier. In the case of snow fences, the objective is to capture and more uniformly distribute the snow behind the fences, which is important for water conservation in regions where snow accounts for much of the precipitation. Snow fences often are placed at critical points on the landscape (e.g., beside roads and highways) where snow tends to accumulate during storms. Terraces can be used to retain water on the land or to carry it away slowly, thereby increasing the potential for water conservation (*see* **WATER CONTROL PRACTICES, *Terrace***). Various type of terraces are available for conserving soil and water. The spacing of plant barriers depends on land slope for controlling erosion by water and on plant height for controlling erosion by wind.

### *Annual Plants*

Almost any plant species that can be grown relatively densely can be used as a barrier. The advantages of using annual plants include easy establishment without the need for special seeding equipment, relatively rapid growth, and readily available seed supplies. In addition, for the next crop they can be "destroyed" during normal land-preparation operations or located elsewhere. The need to reestablish the barrier each year is the main disadvantage of using annual crops.

© 2006 by The Haworth Press, Inc. All rights reserved.
doi:10.1300/5678_03

Another disadvantage is that the barrier plants compete with crop plants for water and plant nutrients.

### *Contour Border Strip*

Contour border strip barriers are strips of a perennial forage crop placed on the contour between strips of annual crops on critical slopes and other vulnerable parts of cultivated fields where erosion cannot be adequately controlled by the annual crops alone or by ordinary strip cropping.

### *Hedge/Hedgerow*

A hedge or hedgerow is composed of a group of plants (e.g., small trees, shrubs, and bushes) growing close to each other in a row that forms a continuous mass of foliage, as in the property line or border between adjacent tracts of land. It may be trimmed periodically or allowed to grow naturally. A hedge or hedgerow can slow the water flow rate across the land, thereby reducing the potential for erosion by water and improving the potential for conserving water. It can also serve as a barrier to reduce the susceptibility of a field to erosion by wind to some distance leeward of the hedge or hedgerow, depending on the height and density of plants.

#### *Contour Hedge/Hedgerow*

A contour hedge or hedgerow is oriented on or close to the contour across a slope, generally on relatively steeply sloping land. Orientation on the contour maximizes the potential for controlling erosion and conserving water, thereby contributing to stabilization of the watershed, preservation and protection of downstream areas, and productivity of the land.

#### *Multiple-Use Hedge/Hedgerow*

A producer can derive a variety of products by using different plants in one or more multiple-use hedges or hedgerows on a given farm or tract of land. For example, trees can provide fuel, poles, posts, timber, shade, environmental protection, and fruits or nuts. When forage plants are included, fodder for livestock becomes available, which then can result in manure becoming available for maintaining soil fertility. Properly designed and maintained multiple-use hedges or hedgerows also generally provide for erosion control and water conservation.

### *Narrow-Strip System*

A narrow-strip barrier system involves strips of crops separated by narrow strips of barrier plants. The goal is to shorten the slope length in the direction of water flow, thereby reducing the erosive potential of the flowing water. Barrier strips for this purpose should consist of sturdy, closely spaced plants to retard water flow downslope. Decreasing the flow rate allows more time for water infiltration, thus providing water conservation benefits. A barrier system also helps control erosion by wind. For maximum effectiveness, the barrier strips should be perpendicular to prevailing winds during the main erosion period and have a porosity of approximately 60 percent. The height of barriers affects the distance downwind that erosion is reduced.

### *Nonvegetative Barriers (Snow Fences)*

Snow fences capture snow and spread it more uniformly on land away from traffic lanes, thus reducing the potential for blocked traffic. Subsequently, the melting snow provides for water storage in soil (water conservation) over a larger area.

### *Perennial Plants*

The requirements for perennial plants are the same as for annual plants, namely, that they form a relatively dense barrier for controlling erosion by water and a relatively tall and somewhat porous barrier for controlling erosion by wind. The main advantage is that the barrier does not need to be reestablished each year. Disadvantages include the need for special seeding equipment for some types of seed, relatively high seed costs, relatively slow growth of some species, and the need to avoid damage to the barrier plants during land-preparation operations. As for annual plants, competition with crop plants for water and plant nutrients is a disadvantage. In addition, some perennial plants are host plants for insects that subsequently infest the crop being grown. In the United States, tall wheatgrass often is used as a perennial barrier plant for controlling erosion by wind and capturing snow in the northern portions of the Great Plains.

### *Shelterbelt*

*See* **BARRIERS,** ***Windbreak.***

### *Ter-Bar*

A ter-bar, or terrace-barrier, is formed by growing barrier-type plants on terrace ridges to help control erosion by wind.

### *Vegetative Barriers*

A vegetative barrier is any barrier consisting of annual or perennial plants. Vegetative barriers help control erosion by water and wind and may provide water conservation benefits.

### *Vetiver Grass*

Vetiver grass is a tropical or subtropical perennial grass that forms a dense vegetative barrier. When planted in rows across the slope of the land at appropriate intervals, such barriers retard downslope water flow and provide for erosion control. Retarding water flow provides more time for water infiltration, thereby enhancing the potential for water conservation. It also traps sediments carried by the water on the upslope side of the barrier, so that the land slope between the barriers tends to decrease with time, tending further to reduce the potential for erosion and increasing the potential for water conservation.

### *Windbreak*

A windbreak is a living barrier consisting of trees and/or shrubs. Such barriers are situated adjacent to fields, farmsteads, and feedlots, for example, to reduce the potential for erosion by wind, protect soil resources, control snow deposition, conserve water, and provide shelter for livestock.

## BEST MANAGEMENT PRACTICES

Best management practices (BMPs) are those recognized to be effective for soil conservation purposes that also provide water quality benefits. They were developed and implemented in the United States as a requirement of the 1977 amendments to the Clean Water Act. They include such practices as cover crops, green manure crops, and strip cropping to control erosion, and soil testing, targeting, and timing of chemical applications (similar to integrated pest management) to prevent the loss of nutrients and pesticides. Personnel from the Natural Resources Conservation Service of the United States Department of Agriculture use BMPs to help individual farmers develop effective soil conservation plans for their farms.

## BRUSH

Brush refers to shrubs and small trees that are not a part of the natural (climax) vegetation for a site or are native but greatly in excess of what is normal. Dense brush on rangeland, for example, strongly competes with desirable forage plants for space, water, light, and plant nutrients, thereby greatly reducing the forage available for grazing animals. Dense brush also makes animal management and control difficult. Brush provides protection against erosion by wind; however, it may not help control erosion by water and, therefore, may not provide water conservation benefits.

### *Control*

Brush control often is desirable to improve range productivity or to reclaim abandoned cropland for crop production. Brush can be managed or controlled by mechanical, chemical, or biological means or by prescribed burning. With chemical and biological control methods, the brushy materials are retained on the land, thereby affording continued protection against erosion, at least by wind. In contrast, control by mechanical means or burning may leave the soil surface bare until other vegetation becomes established. As a result, the potential for erosion may be high when these latter control methods are used. All control methods reduce competition between brush and desirable plants for water, but the amount of water conserved may not be changed unless the newly established plants reduce runoff relative to the amount that occurred when brush covered the land.

### *Matting*

Brush matting is the placement of brushy materials on severely eroded land to minimize the potential for further erosion and to conserve water while grasses and trees are being established. Brushy materials along with wire netting may be placed as a matting along streambanks to control erosion.

# C

## CHEMICALS, AGRICULTURAL (RESTRICTIONS ON USE)

Agricultural chemicals such as fertilizers, herbicides, and insecticides are highly important for successful production in many situations. Some, however, are water soluble and are, therefore, transported by surface waters to streams or reservoirs and by percolating water to aquifers or water tables, which may pollute these water resources. As a result, restrictions have been placed on the use of some agricultural chemicals. Proper use (timely applications, proper amounts, etc.) is essential for avoiding undue restrictions on other chemicals for crop production purposes and, therefore, to achieve satisfactory plant growth to help conserve soil and water resources.

## CLIMATE

Climate refers to the average weather conditions that have prevailed in a location over a period of years.

### *Change*

Climate change, especially global warming, is currently receiving much attention in some parts of the world. If global warming occurs to a significant degree, the impact on soil and water conservation may be positive or negative. If, for example, precipitation increases with

© 2006 by The Haworth Press, Inc. All rights reserved.
doi:10.1300/5678_04

global warming, greater plant growth may occur in regions where growth is now sparse because of limited water. The temperature increase could also result in greater plant growth in regions where growth now is limited by cold temperatures. With greater growth, the potential for erosion by water and wind may be reduced by increased vegetation on the soil surface. Appropriate management of such vegetation could also increase the potential for water conservation. One negative effect of increased precipitation could be greater water erosion of soils that are not adequately protected.

## *Characteristics of Regions*

The climatic characteristics of a region—namely, prevailing temperature, precipitation, wind, pressure, and evaporation—strongly influence the potential for erosion by water and wind and whether water conservation is of major importance. The four main climates (arid, semiarid, subhumid, and humid) occur in tropical, subtropical, temperate, and cold regions. These climates differently influence the erosion potential and especially the need for water conservation in the different regions. For example, water conservation for nonirrigated crop production is most critical under semiarid conditions in tropical regions and progressively less critical in subtropical, temperate, and cold regions. The differences result mainly from the higher temperatures and greater potential evaporation (PE) in the warmer regions, even though precipitation (P) may be the same. As the effectiveness of the precipitation received influences plant production, the potential for erosion by water or wind may follow the same trend because the availability and management of plant cover often determine whether the land can be protected against erosion.

### *Ratio of P to PE <0.20: Arid*

Most arid regions are highly susceptible to erosion by water and wind. The potential probably is greater for erosion by wind than for erosion by water, but erosion by water can be severe when rare intense rainstorms occur. Total precipitation in arid regions is low, often highly erratic, and usually of little or no benefit for crop production. Therefore, the crops are irrigated, but water conservation often is highly important because water supplies for irrigation may be limited and because of competition among agricultural, residential, industrial, and recreational users for available supplies.

*Ratio of P to PE 0.20 to <0.50: Semiarid*

The potential for erosion by water and wind is high in many semiarid regions. Both types of erosion are common, but soil texture (sand, silt, and clay content) and land slope strongly influence the erosion potential on a given tract of land. As more precipitation is received than in arid regions, erosion by water often is a problem on sloping soils, regardless of soil texture. Erosion by wind, on the other hand, usually is more common on sandy than on clayey soils because clayey soils retain water longer near the surface, and surface roughness, which helps control erosion by wind, is more easily produced and maintained on clayey soils. Erosion by wind can, however, be severe on clayey soils when the surface is not protected by plants or adequate roughness, especially when it becomes smooth after rain or when surface soil aggregates have disintegrated under freezing-thawing conditions. Precipitation in semiarid regions generally is adequate for producing some crops without irrigation, but water conservation is highly important for successful dryland (nonirrigated) crop production. Yields of dryland crops generally increase with increased amounts of plant-available water stored in soil at the time a crop is planted, making water conservation (storage in soil) both during the period between crops and during the crop's growing season important.

*Ratio of P to PE 0.50 to <0.75: Subhumid*

The potential is generally much greater for erosion by water than for erosion by wind in subhumid regions. Erosion by wind, however, does occur, especially on unprotected sandy soils. As precipitation is greater and generally more frequent, the emphasis on water conservation is less in subhumid regions than in semiarid regions. Periods without adequate precipitation do occur, however, and water conservation is beneficial for successful crop production without irrigation.

### *Ratio of P to PE >0.75: Humid*

Greater precipitation makes erosion by water the dominant problem in humid regions. Erosion by wind, however, may occur, especially on unprotected sandy soils. In addition, water conservation is beneficial on some soils that are shallow and/or have a low water storage capacity. On such soils, plants may become stressed for water within 5 to 10 days without rain.

### *Drought*

A drought is a continued period during which precipitation is so low that crops either do not mature properly or fail. Limited plant growth may result in inadequate ground cover to control erosion by wind. Then, when precipitation occurs, inadequate ground cover may result in conditions conducive to erosion by water. Water conservation is minimal or nonexistent during droughts; it may even be low when precipitation occurs because of excessive runoff resulting from the poor ground cover.

### *Flash Flood*

A flash flood results from the rapid accumulation of water in streams or low places during a rainstorm. A flash flood may occur in the area where the rain falls or some distance downstream where it has not rained. Flash floods can have disastrous consequences in either case. Flooding normally occurs when heavy rains occur in a given area, and precautions against flooding should be taken. Downstream flash floods may be entirely unexpected and may result in heavy losses of lives and property.

### *Length of Growing Period*

According to the definition given by the Food and Agriculture Organization of the United Nations, the growing period is the number of days during a year when precipitation is greater than 50 percent of potential evapotranspiration plus the number of days needed for plants to use an assumed 100 mm of excess water from precipitation (or less, if not available) stored in the soil. Days on which the temperature is too low for plant growth (usually below 6.5°C) are excluded from the growing period. In general, short growing periods result in relatively little plant growth and cover, which may increase the potential for erosion by water or wind and reduce the potential for water conservation. In contrast, longer growing seasons usually result in greater plant growth and cover, which, properly managed, can reduce the potential for erosion and increase the potential for water conservation.

## *Rain/Other Precipitation*

Rain and other forms of precipitation (snow, sleet, hail) provide the water needed for the production of crops and other vegetation where irrigation is not used and even provide some water for irrigated crops. Precipitation, therefore, has a direct effect on the growth and productivity of crops and other types of vegetation, which, in turn, affects the potential for soil and water conservation.

### *Amount*

Average annual precipitation in a given region has a strong effect on which type of erosion (water and/or wind) will occur and on the importance of water conservation. The average, however, represents the average of amounts received for a number of years and certainly is not the actual amount that will occur in a given year. In general, erosion by water rather than by wind is common in humid regions, except possibly on some soils when land preparation for crops is under way or when land surfaces are disturbed by construction (e.g., road building). Erosion by water and especially erosion by wind are common problems in semiarid and arid regions because plant growth often is too limited to provide adequate cover and crop residues to protect the soil. In addition, water conservation (storage in soil) usually is more important in drier regions because of the generally longer intervals between precipitation events.

### *Annual Distribution*

Precipitation distribution throughout the year strongly affects the crops that can be grown and, therefore, the potential for erosion and the need for water conservation. With a relatively uniform distribution of adequate precipitation, it is usually possible to have one or more adapted crops on the land throughout the year, which may provide adequate protection against erosion. Under such conditions, precipitation may be frequent enough that water conservation is of relatively little importance. When precipitation is highly seasonal, crop growth may be possible only during the rainy season or for one crop after the rainy season, provided adequate water has been stored in the soil to sustain the crop through its growth and reproductive stages. Erosion by water can be a major problem during the period of high precipitation, whereas erosion by wind may be high during the ensuing dry period.

*Excess Precipitation/Unusual Storms*

Large-scale runoff and erosion by water may result from periods of excess precipitation leading to soils being filled with water or from unusually large precipitation events (storms), which generally cause floods because the precipitation rate exceeds the rate at which water can infiltrate the soil.

*Hail*

Hailstones of varying size may occur during some precipitation events (e.g., thunderstorms). Small hailstones usually cause little or no damage to crops or other vegetation and, therefore, have little effect on soil and water conservation. In contrast, large hailstones, especially when numerous, may severely damage or destroy crops and other vegetation, which then increases the potential for erosion and reduces the potential for water conservation. Besides the damage to vegetation, the soil surface may become compacted, thus increasing runoff when subsequent precipitation (rain) occurs. An indirect effect of such events is the reduced crop yield, which lowers the financial income for producers and landowners and possibly their willingness and ability to adequately support or maintain soil and water conservation practices. Melting hailstones may provide some water for conservation purposes.

*Impact*

Precipitation impact on the soil surface, as from rain, sleet, and hail, can greatly influence runoff, water infiltration, and, therefore, the potential for erosion by water and for water conservation. Raindrops impacting the surface disperse soil aggregates, leading to movement of fine soil particles. This results in surface sealing, which reduces infiltration and causes greater runoff, thereby reducing the potential for water conservation. Cover provided by plants and crop residues dissipates the energy of precipitation that strikes the soil surface.

*Intensity*

Precipitation intensity strongly influences the potential for erosion by water and for water conservation. Soil aggregate dispersion owing to raindrop impact and surface sealing usually does not occur during low-intensity storms, when all water may infiltrate the soil, thus avoiding runoff that could cause soil and water losses. In contrast, high-intensity precipitation during rainstorms often results in severe aggregate dispersion, surface sealing, high amounts of runoff, and potentially severe erosion by water unless the surface is adequately protected by plants or crop residues or other erosion control practices are in place. Runoff and erosion may still be high with plant cover on the soil if the precipitation rate greatly exceeds the rate at which water can infiltrate the soil.

*Kinetic Energy*

Kinetic energy is energy in motion, as in the case of falling raindrops, sleet, and hailstones. The total energy in a given precipitation event depends on particle size and terminal velocity and on the total amount of precipitation. The kinetic energy of a precipitation event strongly influences the potential for erosion by water and for water conservation.

*Probability*

The probable amount of precipitation at a certain time of the year has a strong influence, for example, on what type of crop can be grown and when to plant the crop, especially for dryland (nonirrigated) crops.

Average precipitation may also indicate the yield potential of a crop. The probability is based on a record of past precipitation amounts and distribution.

*Return Period*

The return period is the probable time, measured in years, when a rainstorm of a given intensity and duration will recur at a particular site. Calculations of return periods are used in designing various erosion control and water conservation structures and practices (e.g., terraces, waterways, and ponds).

*Snow and Sleet*

Snow, which consists of particles of water vapor that freeze in the upper air and fall to the ground as soft, white, lightweight crystalline flakes, provides much of the water for crop production and other vegetation in northern and southern parts of the world. Snow's light weight results in major irregularities in its deposition when it falls under windy conditions, which then results in uneven water storage in soil when the snow melts. Improved uniformity of deposition can be achieved by keeping standing crop residues on the land during the snowy season. The amount of water stored in soil from melting snow generally increases with increases in the height of crop residues. Other practices to increase snow trapping include the use of vegetative barriers (rows of plants at intervals across the field) and snow plowing, which is a type of snow harvesting that involves forming ridges of snow that trap subsequent snowfall between them. Snow by itself does not result in erosion. Erosion by water, however, may occur when snow thaws, especially when thawing is rapid. Rapid thawing also results in little or no water conservation when soil beneath the snow is frozen. Sleet is frozen rain and does not result in erosion, except possibly when it melts, when the mechanism is the same as for melting snow.

*Temperate versus Tropical*

Precipitation generally is less intense in temperate regions than in tropical regions. As a result, the potential for erosion generally is greater in tropical regions, which, therefore, indicates a greater need for erosion control practices. There are, however, differences in conditions among different tropical and temperate regions (e.g., soils, topography), as a result of which the conditions of a given region must be carefully evaluated to determine which practices are appropriate for that region.

*Timing*

The timing of precipitation has a strong influence on crop yields, especially those of dryland (nonirrigated) crops in subhumid to semiarid regions. Total precipitation during a crop's growing season may be below or above average; however, unless adequate water is available at a critical growth stage (e.g., at early stages for plant development or at the grain-filling stage), yields may be lower than expected based on the total amount of precipitation received. Good amounts of water stored in the soil can improve the potential for favorable yields if precipitation does not occur at the critical growth stage. With poor growth, plants may not provide adequate cover to protect the surface against erosion or to yield water conservation benefits.

*Variability*

As a rule, precipitation variability is significantly greater in dry regions than in humid regions. Plant growth in dry regions, therefore, may be limited to the extent that inadequate cover is available to reduce the potential for erosion. With limited plant growth, the potential for water conservation usually is also decreased. Less variable precipitation in humid regions usually provides the conditions for greater plant growth, which, with proper management, improves conditions for soil and water conservation.

***Temperature***

The prevailing (air and soil) temperature in a given region is a characteristic of the region's climate that is influenced by such factors as latitude, elevation, proximity to mountains or bodies of water (lakes and oceans), and prevailing winds. It strongly influences the potential for soil and water conservation or losses in the region in numerous ways, including its effect on such factors as crops grown, farming systems used, planting dates for crops, evaporation, soil freezing and thawing, and growing season length.

### *Weather*

Weather refers to the general condition of the atmosphere at a particular time and place with regard to such factors as the temperature, moisture condition (rain and snow), cloudiness, wind, evaporation, and humidity. Weather, therefore, has a major influence on the potential for erosion by water or wind and for water conservation. Under some conditions, heavy rains increase the potential for erosion by water and strong winds increase the potential for erosion by wind. The amount of rain and its frequency, temperatures, and wind speeds strongly influence the potential for water conservation.

### *Wind*

For erosion by wind to occur, the wind's velocity at the soil-air interface must exceed the threshold velocity, which is the velocity required to initiate soil movement. Indirectly, wind may also increase the potential for erosion by water and reduce the potential for water conservation. Wind-driven precipitation can increase surface soil aggregate dispersion and, therefore, lead to surface sealing. Surface sealing enhances runoff and reduces infiltration, which may increase erosion by water and reduce water conservation. Factors determining erosion by wind include its speed, direction, and season of greatest speeds relative to soil conditions.

## COMPETITION

With respect to water, competition is the struggle for the available supply—for example, between crops and other plants, between the needs of plants and those of humans or animals, and between agricultural and other users. Competition usually becomes especially critical where the supply of water is limited, thus often making water conservation difficult. Likewise, competition for land among agricultural and other users may affect how the land is managed with respect to soil and water conservation, and competition for financial resources may affect whether soil and water conservation practices are adopted and implemented.

### *Among Sectors of Society*

Competition for water and land among agricultural, residential, industrial, and recreational users is increasing (*see* **LAND, *Change in Use***). Competition for water may result in less irrigation of agricultural land, which can lead to the potential for greater erosion (less plant cover) on the land farmed without irrigation. Depending on the new use of a piece of land and how it is managed, the potential for erosion may increase or decrease. Competition for financial resources among different segments of a society may also determine whether conservation practices are applied to the land. Implementation of some conservation practices is costly and is often supported, at least in part, by funds allocated by the government for conservation programs. Funding for conservation programs may be adequate when government funds are adequate but may be curtailed or eliminated when funds are limited or when other, higher-priority programs require funding from the government.

### *Between Crops and Other Plants*

Competition for water, plant nutrients, light, and space between crop plants and other plants such as weeds, barrier plants, and trees and bushes may lead to poor crop plant growth and yields (*see* **BARRIERS; CROPPING SYSTEM/SEQUENCE, *Alley Cropping***). This may directly influence erosion by resulting in limited plant cover, thereby leading to greater runoff and erosion by water and, under extreme conditions, greater susceptibility to erosion by wind. Indirectly, poor crop yields may lead to financial constraints for the producer or landowner, resulting in a delay or failure to implement or maintain necessary conservation practices. In addition to direct competition for water with crops during the growing season, water use by noncrop plants can seriously hinder or even prevent any water being conserved before a crop is planted.

## CONSERVATION

Conservation, in its broadest sense, is the act or practice of conserving something (protecting it from loss or waste) so that it will be available in the future. With regard to soil and water, conservation is the protection of these resources through the use of appropriate principles and practices that ensure optimum economic and social benefits for humans and the environment.

### *Compliance*

Conservation compliance is the act of adhering to the procedures implemented for conserving soil and water in a given situation. In the United States, the Food Security Act of 1985, as amended in the 1990, 1996, and 2002 Farm Bills, requires that everyone who produces agricultural commodities must protect all cropland classified as highly erodible from excessive erosion. The purpose of the legislation is to remove the incentive to produce annually tilled agricultural commodity crops on highly erodible land unless the land is protected from excessive soil erosion.

### *Cross-Compliance*

Cross-compliance is the mandatory requirement that approved soil conservation practices be used by producers in the United States for them to remain eligible to receive other benefits from the government.

### *Legislation*

Governments in many countries have enacted legislation pertaining to soil conservation. Producers who participate in government programs can receive financial or technical assistance to implement appropriate conservation practices on their land.

### *Programs*

In the United States the Great Plains Conservation Program (GPCP) established in 1956 and the Conservation Reserve Program (CRP) established in 1985 were designed to take highly erodible land out of crop production. To participate in the CRP, for example, farmers contract with the government to idle highly erodible land in return for monetary payments. The contract period is 10 years (some extensions have been granted). In addition to taking land out of production, farmers are required to establish grass or trees on the land. The grass cannot be used for animals, except that some haying and grazing have been allowed under emergency conditions (e.g., prolonged drought). Similar conditions were in place for the GPCP (*see* **GOVERNMENTAL FACTORS, *Farm Programs, Federal***).

### *Tillage*

*See* **TILLAGE, *Conservation Tillage*.**

## CONSTRUCTION ACTIVITIES OTHER THAN AGRICULTURAL

Unless appropriate measures are implemented before construction begins and are maintained until final protective measures are installed, erosion can be a severe problem on construction sites (*see* **LAND, *Change in Use***). Water conservation usually is not of major importance during the construction period, but it may become important after construction ends to maintain vegetative cover on roadsides and landscaped areas surrounding buildings, industrial sites, and so on. Because highway and road construction sites usually are much larger than other construction sites, the discussion here will deal with highway and road construction, but the comments are applicable to other sites in most cases.

### *Cuts*

A cut is the place in a hill that reduces the grade for a highway or road where it crosses that point in the landscape. Erosion may be a problem at a cut site, depending on the earth material at the site (*see* **CONSTRUCTION ACTIVITIES OTHER THAN AGRICULTURAL, *Highways/ Roads,*** *Slope*). Where the soil is suitable, vegetative cover can be established. For cuts with steep side slopes, special provisions usually are required to control erosion or to prevent stones and other materials from falling onto the highway or road.

### *Highways/Roads*

Highway and road construction projects involve a range of activities that often lead to erosion during construction and sometimes after the project is completed. While construction is in progress, vegetation along the intended roadway is eliminated, and even vegetation adjacent to the roadway often is severely disturbed or destroyed in the process of creating suitable construction conditions. When precipitation occurs during the construction period, erosion by water often results because the surfaces are not protected against the action of falling raindrops and flowing water. In some cases, erosion by wind occurs on the unprotected surfaces and along roads traversed by earth-moving equipment.

#### *Alignment*

Alignment of the course of a highway or road across the landscape has a strong influence on potential associated erosion problems. To ensure the fewest problems, the course should be along the crest of the land, thereby avoiding any need for drainage and allowing runoff from the highway or road to be easily discharged to both sides. Where the highway or road must cross steeply sloping land, a diagonal or zigzag course across the slope at an appropriate gradient or cuts in the hills may be required. Under such conditions, appropriate measures to control erosion usually are needed (*see* **CONSTRUCTION ACTIVITIES OTHER THAN AGRICULTURAL, *Cuts***).

*Diversions/Outlets for Runoff Water*

Highway and road construction usually results in the natural channels for water flow across the landscape being altered. Furthermore, it usually results in concentrated water flow adjacent to the highway or road and where the water is allowed to pass under bridges or through conduits beneath the highway or road. Such concentrated flow can cause serious erosion if the diversions and outlets for the runoff are not properly designed and constructed.

*Ground Cover*

In most cases, vegetation is established adjacent to a highway or road as soon as is practical after the construction has been completed. Such vegetation should be carefully managed and allowed to become fully established before being, for example, mowed to control weeds or to aesthetically improve the roadside conditions. Mowing grasses before they become fully established has been observed to seriously hinder their subsequent growth, resulting in severe erosion of the roadside embankment and requiring additional erosion control activities.

*Slope*

Slopes where highways and roads cross the land are highly variable in some regions. Therefore, a thorough knowledge of the region (e.g., precipitation amounts and patterns, soil conditions, adapted grasses) is required to properly design and establish conditions that will minimize the potential for erosion, especially by water, during the project and especially after it has been completed.

# CONTOUR

A contour is an imaginary line on the land surface that connects points at the same elevation.

### *Contour Farming*

Contour farming is a method of farming in which field operations such as plowing and crop planting, cultivating, and possibly harvesting are performed on the contour. Such farming practices can reduce erosion by water and conserve water if appropriate tillage methods are used to retain water on the land.

### *Contour Trenches*

Contour trences (or ditches) are channels for conveying irrigation water or storm runoff water across the slope of the land. They are laid out at a slight grade (not exactly on the contour) so the water will flow at a nonerosive rate to the place to be irrigated or to an appropriate outlet for runoff water.

# CROP

Crop characteristics such as plant leaf type, growth habit, plant spacing, row spacing, and growing season strongly influence the erosion potential of land on which a crop is grown. Close spacing, both within rows and between rows, of plants with a spreading growth habit and large leaves generally provides ground cover quite rapidly after establishment on productive soils and, therefore, readily provides for protection against erosion provided conditions are suitable for rapid or normal growth. In contrast, slow growth resulting from adverse climatic conditions such as cool temperatures and limited precipitation and rapidly drying soil conditions can lead to erosion by water and wind, even for such crops. Erosion by water and wind may occur initially after crop establishment, and conditions conducive to erosion by water may persist throughout the growing season for crops of widely spaced single-stemmed plants with narrow leaves. Widely spaced plants may not protect the soil surface against raindrop impact, which disperses soil aggregates and, in turn, causes soil surface sealing, increased runoff, and limited water conservation.

For crops to provide the most effective erosion control, their growing season should be such that plants are well established and adequately cover the surface when the potential for erosion is greatest. However, this is not possible under some conditions because of prevailing temperatures and precipitation patterns. For example, warm-season crops cannot be established until soil temperatures become warm enough, which may occur later than the main period for wind erosion in some regions (e.g., the western part of Texas in the United States). Even after the crop has been planted and seedlings have emerged, windborne soil particles (sand) sometimes damage or even destroy the seedlings. Cool-season crops may provide for better control of erosion by wind, but the precipitation pattern generally favors warm-season crops in western Texas.

For a discussion of the reasons for growing particular crops, *see* **CROPS**.

### *Litter*

Under field crop production conditions, litter is the chaff, husks, and small stems deposited by harvesting equipment. It also includes fallen leaves and partially decayed residue fragments on the soil surface from previous crops. Litter protects the surface from raindrop impact, thereby reducing the potential for runoff and erosion by water and improving conditions for water conservation.

### *Ratoon*

A ratoon crop is one that regrows from the existing plant after the initial harvest. Such cropping typically occurs for sugarcane and pineapple in tropical regions. Ratoon cropping generally results in a relatively rapid development of surface cover without any land preparation that would disturb the soil, thereby minimizing the potential for soil and water losses.

### *Recropping*

Recropping refers to the practice of growing the same crop on a given tract of land in successive years (or seasons). Under dryland (nonirrigated) conditions, this practice is strongly dependent on water stored in soil in the interval between crops because a fallow period, which normally would be used to increase water storage, is avoided. Under some conditions, erosion by water or wind during a fallow period is a major problem that can be avoided when recropping is practiced.

### *Rotation*

Crop rotation is a production system that involves growing two or more crops on a given tract of land during each cycle of the rotation, which may range from 2 to several years depending on the crops (5- to 8-year rotations are common in some regions). The benefits include improved soil water and nutrient utilization, improved pest control (e.g., of weeds, insects, and plant diseases), better use of crop production resources (machinery and labor), increased potential for some production from crops in a given year (when different phases of the same rotation pattern are carried out on the farm in a given year), and improved soil and water conservation when effective crops for controlling erosion are included in the rotation.

### *Stand Establishment*

The use of high-quality seed planted with proper equipment when conditions are optimum is essential for rapidly establishing a crop at the desired plant population. Such stand establishment minimizes the time during which land is subject to erosion. It also helps to conserve water that becomes available during the crop's growing season.

### *Type Grown*

The type of crop grown can greatly influence soil and water conservation. On relatively flat land, almost any crop may be suitable. On sloping land, however, close-growing crops such as small grains and grasses (planted within and between rows or broadcast planted) usually effectively control erosion and provide for water conservation. (Broadcast planting involves scattering the seeds on the soil surface rather than placing them in the soil with a drill or other planting implement.) In contrast, single-stem plants at relatively wide spacings (within and between rows) on sloping land can lead to greater runoff, potentially greater erosion by water, lower infiltration, and less water conservation. Crops that provide for little or no residue on the soil surface after harvest may result in greater erosion by water or wind, lower water infiltration, and less water conservation.

### *Variety/Cultivar Selection*

A wide range of varieties and cultivars have been developed for many crops, each for a specific region based on such factors as length of growing season, temperature regime, potential water supply (e.g., dryland versus irrigated), and day length. Selecting an adapted variety or cultivar increases the potential for better plant growth, development, and yield, which provides for greater soil cover to reduce the potential for erosion and increase water conservation. The greater yields provide higher income, which may induce a farmer to maintain or implement soil and water conservation practices.

### *Yields*

Satisfactory crop yields are the goal of crop production efforts. Through proper management, including the use of appropriate soil and water conservation practices, the potential for achieving satisfactory crop yields is enhanced. Satisfactory yields mean that conditions remain favorable or may even be enhanced for the producer or landowner to invest in even more effective practices for conserving soil and water. Improved conditions for soil and water conservation may result in greater plant growth and soil cover, thus providing greater yields.

## CROPPING SYSTEM/SEQUENCE

A cropping system or sequence is the established method or plan by which crops are produced. The system used in a given situation is influenced by many factors, including the climate (amount and reliability of precipitation, temperatures), water availability (precipitation and irrigation), soils, crop adaptability, commodity prices, and producer preferences. With regard to soil and water conservation, the goal should be crop production operations that provide for timely crop establishment so that the soil surface is adequately protected during the critical erosion periods.

### *Alley Cropping*

Alley cropping is the practice of growing crops in narrow strips between rows of shrubs or bushes. Trimmings from the shrubs or bushes serve as a mulch on the cropped area. Trimmings from leguminous shrubs or bushes provide nutrients to the crops. Larger branches may be used as fuel for cooking, and some of the leafy materials may be used as feed for animals. Because shrubs or bushes compete with crops for water, alley cropping is best suited to humid regions where precipitation generally is adequate to supply the needs of both the crops and the shrubs or bushes.

### *Continual Cropping*

Continual cropping (often referred to as continuous cropping or annual cropping) is the growing of the same crop on a given tract of land in successive years. Warm- or cool-season crops may be grown. One disadvantage is that a given crop may not have been planted or have become well established when the potential for erosion is greatest. Another is the potential for greater weed and plant insect/disease problems than where, for example, crops are grown in a rotation involving two or more crops.

### *Diversified Cropping*

Diversified cropping is the practice of growing more than one type of crop on a farm in a given year or cropping season. Examples include the combination of warm-season and cool-season crops (one or more of each), two or more warm-season crops, or two or more cool-season crops, each being grown on a different tract of land on the farm in a given year or cropping season. The potential for success is enhanced when livestock are included in the farming operation because the livestock can utilize the grain or forages produced.

### *Double Cropping*

Double cropping is the practice of planting a second crop immediately after harvest of the first crop. It is a form of mixed cropping (*see* **CROPPING SYSTEM/SEQUENCE**, ***Mixed [or Multiple] Cropping***) and is possible where precipitation is adequate or water for irrigation is available and where the growing season is not restricted by low temperatures. Double cropping provides for plants being on the land during much of the year, thereby reducing the potential for erosion and possibly increasing the potential for water conservation.

### *Ecofallow*

Ecofallow is a cropping system designed to control weeds and to conserve water during the period between crops with minimal disturbance of the soil and crop residues on the soil surface. Weeds may be controlled with herbicides or tillage from the harvest of one crop until planting of the next crop. The surface residues protect the soil against erosion by wind and water. They also provide conditions conducive to retarding runoff and enhancing water infiltration, which enhances the potential for water conservation. Provided adequate crop residues are retained on the soil surface, ecofallow is a type of conservation tillage (*see* **TILLAGE,** ***Conservation Tillage***).

### *Ecofarming*

Ecofarming is a reduced tillage cropping system that involves a rotation of winter and summer crops (two crops in 3 years: e.g., winter wheat–fallow–corn or grain sorghum–fallow). In addition to its soil and water conservation benefits (*see* **CROPPING SYSTEM/SEQUENCE, *Ecofallow***), ecofarming reduces the potential for plant diseases that often result from growing a crop continually on a given tract of land.

### *Fallowing*

Fallowing is the practice of allowing cropland to remain idle during a major portion of the season or the entire season when a crop would normally be grown. The land may be tilled or remain untilled during the fallow period. The usual goal is to increase water storage in the soil for the next crop, but weed control, plant nutrient, insect, and plant disease concerns may be involved also. Erosion by water or wind may be a problem during the fallow period when surface cover is low or nonexistent or when other soil conservation practices are not being followed.

### *Flexible Cropping*

A flexible cropping system is one in which the producer chooses to grow crops depending, for example, on prevailing climatic and market conditions. If soil water conditions are favorable soon after harvesting a crop, a second crop may be planted immediately (double cropping) without allowing the land to remain idle (e.g., fallowing or continual cropping). Also, a crop for which the market outlook is favorable may be substituted for a crop for which the outlook is less favorable. Such cropping practices may affect soil and water conservation, depending on the type of crop grown (e.g., close-growing versus more widely spaced plants, erect versus spreading growth habit, and type of residues produced).

### *Integrated Cropping*

Many types of integrated cropping systems are used throughout the world. Some systems involve only one crop; others involve two or more crops. Other systems involve field crops in conjunction with livestock, with the animals being permitted to graze the crop in its early growth stage (e.g., cattle grazing winter wheat in the United States) or to forage on the crop stover after harvest. Alternatively, field crops may be grown in conjunction with tree crops (e.g., trees for fuel, lumber, fruit, and nuts). With appropriate management, including the maintenance of adequate soil cover, soil and water conservation usually can be achieved. Systems involving tree crops are highly effective for controlling erosion by wind, but the trees compete with the field crops for water and nutrients, which may decrease crop yields under some conditions (*see* **CROPPING SYSTEM/SEQUENCE, *Diversified Cropping***).

### *Intercropping*

Intercropping is the practice of growing two or more crops in some combination on a given tract of land in a given year (or similar time period). Examples include field or vegetable crops planted among trees (for fruit, fuel, or lumber), different crops planted within rows (*see* **CROPPING SYSTEM/SEQUENCE, *Mixed [or Multiple] Cropping***), different crops planted in adjacent rows (*see* **CROPPING SYSTEM/SEQUENCE, *Strip Crop Farming***), and a second crop planted where a previously planted crop is approaching maturity (*see* **CROPPING SYSTEM/SEQUENCE, *Relay Cropping***). Intercropping provides for virtually continuous plant cover on the land if year-round plant growth is possible, which benefits soil and water conservation. Where growing seasons are limited by water availability, temperature extremes, or other factors, appropriate management practices may be required to achieve soil and water conservation.

### *Irrigated/Dryland Cropping*

Where water for irrigation is limited and some water is available from precipitation, an irrigated/dryland cropping system can lead to successful crop production under the right conditions. In a given time period (e.g., one year), the crop on one area of land is irrigated while a second crop on another area is grown without irrigation (dryland). In the following period, the crops are switched, with the dryland crop possibly obtaining some water remaining in the soil after harvest of the previously irrigated crop. With proper management, some additional soil water storage may have occurred, which also will benefit the dryland crop. Erosion control also is possible with an irrigated/dryland system.

### *Ley Cropping*

Ley cropping or farming is the practice of growing a grass or a legume in a temporary pasture in rotation with a crop. After some interval, the ley is cropped and a grass or legume is established on the previously cropped area. Erosion control and water conservation benefits occur in the ley owing to stabilization of the soil surface, interception of raindrops, and reduction in runoff rate and amount. Reduced runoff also increases the potential for water conservation. The grass or legume usually is grazed by animals, which helps increase the fertility level in the pasture. After the ley is plowed, the improved soil structure and nutrient level remain for several years.

### *LID System*

The LID (limited irrigation—dryland) system was developed in the southern Great Plains of the United States, where water for irrigation is limited and where dryland (nonirrigated) crops can be grown. It was assumed that a given amount of irrigation water would be available on a regular basis and that it would be applied to a crop when it became available. The upper end of the field was managed for irrigation (appropriate plant populations and fertilizer applications) whereas the lower portion was managed as dryland (lower plant populations). The land was managed for furrow irrigations. Water was applied in small enough quantities for none to flow to the lower end of the field under normal conditions. However, if rain fell shortly before or at the time of irrigation, water advanced farther downslope, thus possibly irrigating the previously dryland part of the field and potentially increasing yields. One goal achieved using the LID system was to improve overall water use efficiency.

### *Mixed (or Multiple) Cropping*

Mixed (or multiple) cropping is the practice of growing two or more types of plants in close association on a given tract of land. It usually provides for production by one crop even when the other fails as a result of weather conditions, insects, or plant diseases, thereby reducing the level of risk for the producer. Mixed cropping is commonly used by producers with limited land areas, and it involves mostly hand labor. Some products may be produced for sale, but the farm must first provide the food needed by the farmer's family. With a variety of crops on a given tract of land, plant cover generally is adequate to protect the soil against erosion by water, and erosion by wind usually is not a problem. Good plant cover also increases the potential for water conservation.

### *Monoculture*

Monoculture is the growing of the same crop on the same land year after year (*see* **CROPPING SYSTEM/SEQUENCE, *Continual Cropping***).

### *Opportunity Cropping*

Opportunity cropping is the practice of establishing a crop whenever conditions become favorable for growing it under a cropping plan or sequence different from that commonly used on a given tract of land. As an example, the winter wheat–fallow–grain sorghum–fallow rotation often used in the southern Great Plains of the United States involves a fallow period of approximately 10 to 11 months between successive crops. For this rotation, one goal is to conserve water during the fallow to provide better conditions for the establishment and growth of the next crop. Sometimes, however, adequate precipitation occurs late in the growing season or soon after harvest of either crop to provide conditions favorable for the establishment and growth of the next crop without going through the fallow period.

Under such conditions, using opportunity cropping, sorghum could be immediately grown after wheat or vice versa, or an alternate crop could be grown to utilize the water that has become available. Opportunity cropping alters the planned rotation but can improve water use and conservation and provide plant cover on the land, which provides protection against erosion by water and wind. One possible constraint on opportunity cropping is the use of soil-applied herbicides. Sufficient residuals of a herbicide used for one crop may remain in the soil to prevent growth of a different crop planted soon after the first crop (without allowing time for the breakdown of the herbicide).

### *Relay Cropping*

Relay cropping is the practice of interseeding a second crop into the first crop well in advance of harvesting the first crop. It is a form of mixed cropping that provides for production of a second crop where the growing season is too short for double cropping (*see* **CROPPING SYSTEM/SEQUENCE,** ***Double Cropping; Mixed [or Multiple] Cropping***). As for double cropping, adequate precipitation or irrigation water is required for relay cropping. Because plants cover the land during much of the year, relay cropping provides erosion control and possibly water conservation benefits.

### *Runoff Cropping*

A runoff cropping or farming system involves capturing runoff water from compacted or specially treated areas and using that water to produce crops, usually on smaller areas. This system is used where average precipitation is not adequate to produce a crop without additional water and where water for irrigation from an outside source (stream, reservoir, or aquifer) is not available. The runoff water can be stored in ponds or tanks, for irrigation as needed, or in the soil where the crop is to be planted. Crops may be typical field crops or individual plants such as a fruit tree or a shrub. Runoff cropping was used for water conservation purposes to enhance crop production in a dry region (Israel) many years ago.

### *Shifting Cultivation*

Shifting cultivation is a cropping system that involves clearing the land, burning the debris, and growing crops for a relatively short period (of the order of several years). When production becomes sufficiently poor, a new area is cleared and the original area is abandoned to allow native vegetation to regrow and restore soil nutrients. Shifting cultivation often is used on sloping lands without provision for soil and water conservation. Increasing population pressures in many cases result in recropping occurring more frequently, which increases the potential for soil and water losses.

### *Strip Crop Farming*

Strip crop farming involves growing different crops in adjacent rows, either at the same time or at different times during the production period. It is similar to mixed cropping in that it provides for more than one crop, which reduces the level of risk for the producer by enhancing the potential for production by one crop if the other crop fails (*see* **CROPPING SYSTEM/SEQUENCE, *Mixed [or Multiple] Cropping***). It also reduces the potential for erosion because some plant cover usually is present during much of the year. Strip cropping can also be used with only one crop (e.g., wheat), essentially with half the field being devoted to the crop in any given production cycle and the other half being fallowed in preparation for the crop in the following cycle.

#### *Buffer (or Spreader) Strip*

A buffer strip is an area above or below a cultivated field or between fields that is not used as a part of the regular farm rotation. It usually is of variable width and is planted to grass or other erosion-resistant plants. Whether the vegetation is harvested depends on the value of the forage relative to the main products on the farm. The objective is to minimize the erosiveness of water flowing onto the field or adjacent fields and to trap sediments that could be carried downslope into streams.

### *Contour Strip Cropping*

Contour strip cropping involves orienting the strips of the protective crop and other crops across the slope of the land, which reduces the rate of water flow downslope, decreases the potential for erosion, and improves conditions for water conservation. Areas between cropped strips usually are of nonuniform width throughout the length of the field, which results in correction or headland strips being needed for the soil-conserving crop.

### *Correction Strip*

A correction strip is the irregular area between strips of crop rows positioned between, for example, adjacent terraces that are not separated by a uniform distance for their entire length.

### *Field Strip Cropping*

Field strip cropping is similar to contour strip cropping in that the strips are oriented across the slope, but in this case not necessarily on the exact contour of the land. It is best suited for fields with a uniform slope. Crops on the different strips may be alternated in successive growing seasons.

### *Filter (Vegetative) Strip*

A filter strip is a narrow area of vegetation adjacent to streams that is used to filter out sediments, nutrients, organisms, organic materials, and chemicals from runoff and waste water before it enters the stream. Filter strips also slow the rate of flow and thereby help to control erosion where the water enters the stream. Filter strips have no effect on the potential for upslope erosion.

### *Headland/Turn/Border Strip*

A headland, turn, or border strip is a small, irregular field area between adjacent strips of row crops in a terraced field. Such areas should be planted to grass when row crops are grown but can be seeded to small grain or hay crops when the entire field is used for such crops. A headland planted to grass can serve as a roadway from one strip to another, a grassed waterway, or an outlet for a terrace or diversion ditch.

*Meadow/Pasture Strip*

A meadow or pasture strip is a vegetative area along which runoff from terraces or diversion ditches, for example, is allowed to flow downslope to an appropriate outlet. Such areas usually are shallow and relatively wide, with grass being used as the stabilizer. They are considered to be either meadow or pasture based on the use made of the forage.

*Parallel Strip Cropping*

In contrast to contour strip cropping, parallel strip cropping involves orienting the strips of the main crop with a uniform distance between them throughout the length of the field. As a result, the strips of soil-conserving crops are also of uniform width, and correction or headland strips are not needed. Parallel strip cropping is most appropriate for areas with a relatively uniform and slight slope.

*Permanent Strip*

A permanent strip is one that remains in place from year to year rather than being cropped in rotation with other crops in the field. Permanent strips usually have perennial vegetation established on them and frequently are used on irregular areas, for correction strips, or on critical slopes or other odd portions of fields where maintaining the regular rotation of strip cropping is difficult.

*Wind Strip Cropping*

For wind erosion control, straight parallel strips planted to regular farm crops are oriented perpendicular to the direction of prevailing winds regardless of the slope of the land. This orientation reduces the distance between the strips where the wind velocity at the surface is above the threshold value needed to initiate soil movement, thus reducing the potential for erosion by wind.

# CROPS

The diverse characteristics of different plant species strongly influence the potential for soil and water conservation. Under this heading, the emphasis is on the reason for using certain crops rather than on the crop itself, as discussed under the **CROP** heading.

### *Canopies*

*See* **CROPS, *Growing Crops.***

### *Catch Crops*

Catch crops are those planted when the crop originally planned could not be established because of unfavorable conditions (e.g., the soil being too dry or too wet) or when the original crop was destroyed (e.g., by hail, insects, or disease) and it is too late to replant the original crop. Catch crops provide for erosion control (as compared with leaving the soil bare and exposed to water and wind) and may provide income for the producer.

### *Companion Crops*

A companion crop is one grown with another crop. Usually, one of the crops grows and matures more quickly and is harvested before the slower-growing second crop, which requires more space for its development. A companion crop may also provide shelter for a second crop planted later.

## *Cover Crops*

Cover crops usually are grown to protect the soil when the potential for erosion is greatest. They may be grown during the time between successive annual crops (primary crops for the given situation) or grown continually on sites where the potential for erosion by water or wind is high. Some of the plant material may be harvested as forage for animals or for other purposes. Cover crops require water for growth and may, therefore, result in less soil water being available for the next crop, especially in regions where water for crops is limited (e.g., semiarid and possibly subhumid regions). When growth is terminated early and precipitation is timely and adequate, growing cover crops may provide water conservation benefits, even in drier regions and especially in more humid regions.

### *Annual Cover Crops*

Annual cover crops are those that are established each year after harvest of the primary crop. Their growth often is terminated before they reach maturity. Residues of annual cover crop plants may be retained on the soil surface to protect primary crop seedlings when the erosion period extends into the growing season of the primary crop. Any crop that grows well during the off-season of primary crops or that provides adequate plant cover during the period of greatest potential for erosion can be used as an annual cover crop.

### *Permanent Cover Crops*

Permanent cover crops are used for long-term protection of highly erosive sites. They are maintained in place from season to season and their main growing season can be at any time of the year. Because the crop's growth is not terminated, the plant materials remain on the soil surface (except when the forage is harvested for other purposes), which promotes water conservation and provides protection against erosion, even if the plants are dormant when the potential for erosion is high. Even when some of the material is harvested, adequate materials usually remain to provide protection against soil and water losses.

### *Forage Crops*

Forage crops are those that are grown as feed for animals, either for grazing in the field or for making silage or hay. Many forage crops are planted at close spacings within and between rows, and they usually provide excellent erosion control and water conservation benefits during their growing season. Excessive grazing or improper management after removing the crop for silage or hay can result in the land becoming susceptible to erosion by water or wind and not conducive to water conservation.

### *Grass*

Land planted to grass usually provides the greatest protection against erosion, both by water and by wind, especially if the grass is a perennial type and overgrazing is avoided. Grass cover also provides suitable conditions for water conservation.

### *Green Manure Crops*

Green manure crops grown during the off-season of primary crops usually are plowed under to improve the fertility and physical condition of the soil. Because such crops provide cover when the land might otherwise be bare during the interval between crops, they offer protection against erosion by water and wind.

### *Growing Crops*

Actively growing crops provide protection against erosion by water if their leaves (the plant canopy) protect the soil from falling raindrops that could disperse soil aggregates, cause soil surface sealing, and increase runoff. At close plant spacings, growing crops also impede surface water flow, thereby protecting the land against erosion by water. Well-established growing plants provide excellent control of erosion by wind.

### *Legumes*

Legumes are one of the most important and widely distributed plant families and include such food and forage species as peas, beans, peanuts, clovers, alfalfa, lespedezas, vetches, and kudzu. A number of tree and shrub species are also legumes. Most legumes fix nitrogen in association with appropriate organisms in the soil. Therefore, in addition to legumes helping conserve soil and water while they are grown, some nitrogen usually is supplied for the next crop, which may result in it growing better and providing soil and water conservation benefits.

### *Nurse Crops*

*See* **CROPS, *Companion Crops.***

### *Prices of Different Crops, Relative*

When several crops are adaptable to a given region, the crop generally grown is the one resulting in the greatest net financial return to the producer or landowner. This crop may not, however, provide the greatest soil and/or water conservation benefits.

### *Row Crops*

Row crops are those planted in rows spaced wide enough apart to allow cultural operations (e.g., cultivation) during the growing season. To control erosion by water and for water conservation, rows should be close to the contour of the land (across the slope). Row direction should be perpendicular to the prevailing wind direction for greatest protection against erosion by wind.

### *Short-Season Crops*

Crops with a short growing season usually require less water than those with a longer growing season. More water may, therefore, be conserved for a subsequent crop by growing a short-season instead of a long-season crop. One possible disadvantage of growing a short-season crop is that the land may be idle for a relatively long part of the year, thus leaving it subject to erosion by water or wind.

### *Soil-Conserving Crops*

Soil-conserving crops are those with characteristics that provide protection against erosion (*see* **CROP, *Type Grown***) and that help maintain or enhance soil fertility for better growth of subsequent crops. Crops that promote improvements in soil conditions such as aggregation, porosity, and organic matter content, which enhance water infiltration and reduce runoff, also help conserve the soil.

### *Soil-Depleting Crops*

Soil-depleting crops are the opposite of soil-conserving crops. Long-term growing of such crops may degrade the soil to the point where plants grow so poorly they provide little or no protection against erosion and water conservation becomes extremely difficult.

### *Winter Crops*

In some regions, the potential for erosion is greatest during the winter. Although residues of warm-season crops can be managed for soil and water conservation purposes in such regions, greater erosion control generally is possible by growing well-adapted cool-season crops.

# D

## DESERTIFICATION

Desertification is the process of land degradation in arid, semiarid, and dry subhumid areas that turns productive lands into nonproductive deserts. Such degradation results from various factors, including climatic variations and human activities such as allowing overgrazing by animals, cutting of trees and vegetation for fuel and other purposes, use of cultivation practices that induce erosion, cultivation of marginal lands, and improper water management leading to salinization of irrigated lands. Drought generally accelerates desertification in a given region. As desertification progresses, fewer plant materials are produced, which, in turn, tends to accelerate desertification, increases the potential for erosion by water and wind, and reduces the potential for water conservation.

## DESERT TORRENT

A desert torrent results from a rainstorm of the cloudburst type. In desert regions, ground cover is low owing to the sparse vegetation. Therefore, when heavy rains occur, there is little protection against rapid flow of water from the landscape. The resulting torrents cause erosion and may damage any structures in their path. Because of the large amount of runoff, water conservation under such conditions is limited.

© 2006 by The Haworth Press, Inc. All rights reserved.
doi:10.1300/5678_05

## DIRECT SEEDING

*See* **TILLAGE,** ***No-Tillage.***

## DRILL/PLANTER TYPE

*See* **PLANTING/SEEDING,** ***Methods,*** *Drills.*

## DRYLAND FARMING

*See* **AGRICULTURE,** ***Dryland Agriculture***.

# E

## EARTHWORMS

Earthworms and other soil organisms (e.g., insects, spiders) may have positive or negative effects on soil and water conservation. Channels produced by such organisms result in rapid water infiltration into soils, which reduces the potential for runoff and, consequently, for erosion. Rapid infiltration is important also for water conservation. Such benefits result also from improved soil physical conditions attributable to the activities of these. Increased erosion by water may occur, however, when earthworm castings deposited on the soil surface are carried away in runoff during major rainstorms.

## ENVIRONMENT

Environment is the sum of the external factors acting on an organism or community and thereby influencing its development or existence. Many environmental factors affect soil and water conservation, including the physical conditions at a given site (e.g., climate, landscape features), societal education levels, social interest groups, the monetary costs and benefits of using conservation practices, and health issues. Environmental issues are becoming increasingly important in terms of the need to use appropriate measures to conserve soil and water.

© 2006 by The Haworth Press, Inc. All rights reserved.
doi:10.1300/5678_06

# ERODED SITES/CONDITIONS

### *Arroyo*

An arroyo is a gully found in areas of unconsolidated alluvium. The walls of arroyos are mostly vertical and water flow usually occurs only during a precipitation event and for a relatively short time afterward. Flowing water may cause erosion (collapse) of the walls, thus reducing the amount of land adjacent to the arroyos and carrying the sediments downstream.

### *Blowout Area*

A blowout area is a site on the landscape, often on rangeland, that is devoid of vegetation (mainly grasses). Erosion by wind has resulted in a depression in the landscape and erosion continues to be a problem unless remedial actions are taken.

### *Coastal Erosion*

Coastal erosion results from the interaction between the beach and the ocean and is influenced also by human activities. Wind, waves, and ocean currents are among the forces that affect coastal erosion. For example, winter storms and sometimes warm-season storms (hurricanes) may remove significant amounts of sand and create steep and narrow beaches. In contrast, gentle waves in summer return the sand and create wider beaches with gentler slopes. Coastal erosion control is of major importance in many coastal communities.

### *Dune*

A dune is an accumulation of sand and related materials in a ridge or mound as a result of erosion by wind.

### *Creep*

Dune creep is the movement downwind of a dune owing to the transport of windborne sand particles. The creep may be in any direction at a given time, but usually is predominantly in the direction toward which the prevailing winds blow in a given region. With continued creep, land downwind of the dune is lost for crop production.

### *Stabilization*

Dune stabilization can be achieved by erecting "fences" that diminish the wind force at the dune surface, treating the dune surface with materials that stabilize the surface (emulsions), or establishing suitable plants at the base of the dune. Although precipitation usually is low in regions where dunes develop, the water-holding capacity of the soil (mostly sand) in dunes is low. Consequently, even low precipitation in such regions may result in adequate water at the base of dunes to support plant growth, thus helping to stabilize the dunes.

## ***Gully***

A gully is the consequence of concentrated water causing soil to be eroded from the place where the flow occurs. Water commonly flows in an existing gully only during and immediately after a heavy rain or when snow melts. A gully, in contrast to a rill, generally is sufficiently deep that it cannot be obliterated by normal tillage operations. In contrast to an arroyo, which forms in alluvial soils, a gully can develop in any type of soil material (*see* **ERODED SITES/CONDITIONS, *Arroyo***).

### *Control*

Gully control halts further erosion in a gully or remedies the erosion that resulted in the formation of the gully. For effective gully control, factors such as the economics of different control measures, control principles for the given situation, prevailing climatic conditions (e.g., rainfall patterns and temperatures, which affect vegetation establishment and growth), suitability of different control structures, timeliness of the control operation, and suitability of different vegetative materials for controlling erosion must be considered.

*Plug*

*See* **WATER CONTROL PRACTICES,** ***Dams,*** *Check Dam.*

***Localized Scour***

*See* **EROSION,** ***Scour Erosion.***

***Piping***

Piping is the removal of soil by subsurface water flow in channels or pipes developed by seepage water. Water flow in soil cracks, animal and insect burrows, and decayed-root channels may also result in piping.

***Scald Area***

A scald area is a relatively small, highly eroded area where water infiltration is low and where major remedial measures are needed to restore the productivity of the land. Possible remedial measures include ripping, subsoiling, disking, and spike-tooth harrowing.

***Surface Drift***

Surface drift is the movement of soil (sand) by wind from one area (dune or field) to an adjacent area, where it may be trapped in a roughened field area, in a growing crop, or in crop residues. Such movement is at or very near the soil surface because the sand particles are not carried by the wind but skip or bounce along the surface (*see* **EROSION,** ***Forms of Erosion,*** *Saltation*).

# ERODIBILITY

Erodibility is the susceptibility or vulnerability of a soil to erosion. It is influenced by the soil's inherent characteristics, namely, its chemical, physical, and mechanical composition; by the topographical features of the land, mainly its slope; and by how the soil is managed.

# EROSION

Erosion is the wearing away of the land surface through the action of running water, wind, ice, and other geological agents, including the force of gravity, which causes gravitational creep. Human activities result in some types of erosion. Detachment and removal of soil and rock particles are involved in the erosion process.

### *Abnormal/Accelerated Erosion*

Abnormal erosion is faster than is normal for the prevailing conditions. It results primarily from the activities of humans and, in some cases, from animal actions (e.g., making trails, burrows) and uncontrollable disasters (e.g., fires and floods).

### *Animal-Track Erosion*

Animal-track erosion is mainly due to water flowing in the tracks (paths or trails) formed by animals repeatedly walking along the same route across the landscape to feeding areas, water sources, and so on.

### *Badland Erosion*

Badland erosion results in grooving or shelving of the landscape because of unequal removal of erodible materials from exposed strata that have varying resistance to erosion. Such erosion often results in short, steep gullies with peculiar patterns such as those found in many parts of the western United States.

### *Channel Cutting*

Water flowing downslope follows the path of least resistance. As a result, flow usually is concentrated in low places in the landscape. Such concentrated flow may lead to the cutting of a channel (gully or ditch) through which water flows when subsequent runoff occurs. This erosion process contributes sediments to streams and downstream sites.

### *Critical Slope Erosion*

Erosion on critical slopes occurs more rapidly than on lesser slopes. Generally, practices for controlling erosion on critical slopes differ from those required on noncritical slopes. Permanent vegetation such as a perennial forage crop rather than trees or other woody species usually provides effective control.

### *Fertility Erosion*

Fertility erosion describes plant nutrient losses caused by erosion by water or wind. The nutrients lost can be comparable in amount to those removed by crops harvested from the land. Nutrients lost owing to erosion by water may be dissolved in water or adhere to soil particles transported by the water. Losses owing to erosion by wind usually are nutrients adhering to soil particles, but may include part of the soil organic materials. Under some conditions, nutrients being lost from one tract of land may result in nutrient enrichment on another tract if the water- or wind-transported materials are captured and become concentrated.

### *Forms of Erosion*

Erosion by water and wind occur under many conditions.

*Abrasion*

Abrasion is the wearing away of a soil owing to the friction that occurs when soil particles carried by water or wind come into contact with soil. Abrasion also is caused by glaciers, falling rocks, or other shifting materials (e.g., landslides).

*Deflation*

Deflation is the sorting, lifting, and removal of dry, loose, fine-grained particles from the mass of soil by turbulent eddies of wind. In many cases, it also reduces the nutrient level of the soil, thereby potentially reducing its productivity unless nutrients are applied.

*Detrusion*

*See* **EROSION, *Forms of Erosion,*** *Abrasion.*

*Efflation*

Efflation is the carrying away in suspension of very fine soil particles removed from the soil mass by wind.

*Effluxion*

Effluxion is the downwind movement of intermediate-size soil particles owing to the bouncing action called saltation caused by wind.

*Extrusion*

Extrusion is the rolling away of large soil particles during erosion by wind.

*Saltation*

Saltation is soil particle movement by skipping or bouncing on the soil surface caused by wind. It also occurs along the bed of a stream in flowing water.

#### *Suspension*

Suspension is the containment or support of soil particles or aggregates in water or air that allows them to be transported for relatively long distances by water or wind.

### *Geological (Normal) Erosion*

Geological erosion is the normal or natural erosion resulting from the action of erosive processes that occur over long periods. Examples of geological erosion are the wearing down of mountains and the formation of flood and coastal plains. Geological erosion also removes topsoil from undisturbed landscapes at about the rate it is formed under prevailing climatic conditions and deposits it at another location. Geological erosion is also known as the "geological norm of erosion."

### *Gully Erosion*

Gully erosion results from the concentrated flow of accumulated water in narrow channels. The flowing water removes soil from channels to depths usually greater than 0.3 m (as distinct from rill erosion).

#### *U-Shaped Gully*

U-shaped gullies develop in soils and subsoils that are friable and easily cut by flowing water. The sides of the gullies are vertical as a result of the collapse of the banks owing to flowing water undercutting the banks.

#### *V-Shaped Gully*

V-shaped gullies develop in soils with a texture that is relatively more resistant to erosion than that of soils where U-shaped gullies develop—often a heavy clay underlain by a clay. V-shaped gullies usually develop more slowly than U-shaped gullies under similar conditions.

### *Human Influences*

Many human activities have a major impact on the potential for erosion on a given tract of land. Included among these are the management of land for crop production (tillage methods, crops grown, and crop residue management practices), animal production (animal populations and grazing intensity), urban developments (e.g., housing, shopping centers, utilities), road construction, and sporting events (off-road vehicular traffic).

### *Irrigation Erosion*

Erosion caused by irrigation is most likely to occur where the water application rate is relatively high, as with some furrow irrigation. Irrigation erosion is also possible with sprinkler irrigation when the water application rate exceeds the rate of infiltration into the soil.

### *Logging Tracks/Roads*

Logging operations often involve the development of roads on relatively steep slopes to gain access to remove the logs. In addition, moving the logs from the place of cutting to the point of loading onto transport vehicles may involve dragging them across the land. Both factors often result in highly erosive soil conditions.

### *Natural/Normal Erosion*

*See* **EROSION,** ***Geological (Normal) Erosion.***

### *Pedestal Erosion*

Pedestal erosion occurs when the highly erodible soil material is removed from the landscape, except where it is protected by tree roots, stones, or other objects on the soil surface. The erosion is mainly due to raindrop impact and splash action because there is little or no undercutting by flowing water at the base of the pedestal.

### *Pinnacle Erosion*

Pinnacle erosion results in high pinnacles along the sides and in the bottom of gullies formed in highly erodible soils. The gullies have deep, vertical cuts in the sides where erosion is rapid. As erosion progresses in adjacent cuts, they join and isolate the pinnacles. The pinnacles often contain an erosion-resistant layer of stones or gravel, making them similar to pedestals (*see* **EROSION,** ***Pedestal Erosion***).

### *Pothole Erosion*

Pothole erosion describes the type of erosion that results in a corkscrew-like "curvature" in the bottoms of gullies under some conditions.

### *Puddle Erosion*

Puddle erosion describes the physical deterioration of soil structure without net loss of soil from the system because it occurs in a puddle. The structural breakdown is caused by rain washing the fine materials into the depression, resulting in a structureless soil with a "choked" surface and low productivity.

### *Rare-Storm Erosion*

Rare-storm erosion results from intense storms that have an expected return frequency of many years (*see* **EROSION,** ***Abnormal/ Accelerated Erosion***).

### *Rill Erosion*

Rill erosion results in numerous small channels caused by intermittent water flow, as during or immediately after a rain or when snow melts. Rills usually are several centimeters deep with relatively steep sides and occur most frequently on recently tilled land. They do not interfere with, and can be eliminated by, normal cultural operations.

### *Roadside Erosion*

Several types of erosion occur along roadsides. Because road surfaces are mostly impermeable to water, especially when they are paved, runoff from roads results in increased water flow in roadside ditches, which may cause erosion if the ditches are not properly designed to accommodate the increased water flow. Because roads often are built across natural drainage ways, runoff from adjacent land is concentrated and diverted through culverts or pipes under the road. As a result, the greater-than-normal flow of water through these structures may cause erosion if adequate provisions have not been made. Another type of roadside erosion occurs where the land has been cut during road construction to provide for lower grades along the roadway. Such cuts usually result in steep or relatively steep slopes beside the roadway on which it is difficult to establish vegetation that would help control erosion.

### *Rock Erosion*

Rock erosion is the wearing away through abrasion of consolidated rocks on which there is little or no soil material. Such erosion is generally considered to be **GEOLOGICAL (NORMAL) EROSION**.

### *Sand Injury to Plants*

Sand injury occurs when windborne sand particles strike plants during a period of erosion by wind. Young seedlings may be destroyed.

### *Scour Erosion*

Scour erosion occurs when sediment-laden water flows over certain types of dams or in meandering streams with curves of short radius. Sediment-laden water is more erosive that clear water.

### *Sheet Erosion*

Sheet erosion is the removal of a fairly uniform, thin layer of soil from the land caused by water flowing across the surface.

### *Shore/Wave Erosion*

Erosion of reservoir, sea, and ocean shores is caused by the action of waves resulting from wind, boat traffic, and, for seas and oceans, the rise and fall of tides. Some shore erosion may result from water flowing into the body of water from the adjacent land.

### *Soil Phase*

A soil phase is a subdivision of a soil series that is not sufficiently different to be classified as a separate series (*see* **EROSION, *Soil Series***). Susceptibility to erosion may differ for different phases of the same series because of texture, surface slope, past erosion, stone content, and other factors.

### *Soil Series*

A soil series is a group of soils with horizons that are similar in terms of differentiating characteristics and arrangement in the profile, except for the texture of the surface horizon. The susceptibility of different soil series to erosion by water or wind is influenced by such factors as surface texture, surface slope, profile depth, profile horizon thickness, and underlying horizon texture.

### *Soil Type*

Soil type is an obsolete term in the United States. It was a classification of a soil series based on surface soil texture. Soil type is now considered to be a kind of soil phase (*see* **EROSION, *Soil Phase***).

### *Splash Erosion*

Splash erosion is the loosening and splattering of small soil particles caused by raindrops impacting a wet soil surface. The particles may be removed when runoff occurs.

### *Stepped Crescent*

Stepped crescents are shallow crescent-shaped scars resulting from slow downslope creep or mass movement of soil. Such movement may occur beneath sod at the surface. Contributing to the formation of the stepped appearance is the vertical or almost vertical "sinking" of soil within the crescent.

### *Streambank Erosion*

Streambank erosion is the scouring of soil materials and cutting of streambanks by water flowing in streams.

### *Streambed Erosion*

Streambed erosion is the scouring of soil material and cutting of the channel bottom by water flowing in streams.

### *Structural Erosion*

Structural erosion results from large masses of soil falling into gullies from the sides after they have been undercut by water flowing in the gully.

### *Tillage Erosion*

Tillage erosion is the downslope movement of soil caused by the operation of tillage equipment. It is influenced by tillage factors and landscape characteristics. Tillage factors include depth of tillage, speed of operation, and length, width, and type of tillage implement (shape and arrangement of components). Important landscape characteristics include slope curve and gradient: tillage erosion is greatest on convex slopes. Soil properties such as bulk density, water content, and ability to resist displacement also affect the degree of tillage erosion that occurs.

### *Tunnel Erosion*

Tunnel erosion results from water flowing into rodent burrows, channels remaining from decayed plant roots, or soil structural fissures in the landscape. Continued water flow into such openings causes their enlargement and development of the tunnels.

### *Urban Areas*

Erosion under urban conditions often is associated with the development of new residential, commercial, or industrial areas. Erosion by water and wind may occur and is most common after initial land clearing and before the areas are seeded to grass, landscaped, or protected against erosion by other methods.

### *Vertical Erosion*

Vertical erosion is the downward movement of fine clay particles through porous sand or gravel. The particles are carried by water and accumulate in a less permeable zone lower in the soil profile. Such erosion may result in the surface soil becoming less fertile and the zone of accumulation becoming less permeable to water and plant roots.

### *Wash Erosion*

Wash erosion is a term for the erosion caused by water flowing across the soil surface. It is used to distinguish such erosion from splash erosion, which is caused by raindrops impacting a wet soil surface.

### *Waterfall Erosion*

Waterfall erosion occurs where water falls almost vertically into previously formed gullies, trenches, or other features marked by an abrupt vertical or nearly vertical change in elevation of the landscape. Waterfall erosion contributes to gully expansion by undercutting the subsurface layers, mainly at the head of the gullies. It also results in lateral expansion of gullies and the formation of tributaries of the main gullies.

## EROSION PREDICTION

A number of models (equations) have been developed for the prediction of soil erosion by water or wind. Some have wide applicability; others have rather limited applicability. Factors considered in the models include various aspects of the climate, soil, crops (or other plants or plant materials), landscape features, and management options. Further discussion of the different models is beyond the scope of this book. Readers interested in erosion models or equations can find information on the Internet and in numerous publications. Models or equations that have generally wide applicability and that often are used or referred to in the literature include

AGNPS—Agricultural Nonpoint Source Pollution Model
ANSWERS—Areal Nonpoint Source Watershed Environment Response Simulator
CREAMS—Chemicals, Runoff, and Erosion from Agricultural Management Systems
EPIC—Erosion/Productivity Impact Calculator
RUSLE—Revised Universal Soil Loss Equation
RWEQ—Revised Wind Erosion Equation
SWRRB—Simulator for Water Resources in Rural Basins
TEAM—Texas Erosion Analysis Model
USLE—Universal Soil Loss Equation
WEAM—Wind Erosion Assessment Model
WEELS—Wind Erosion on European Soils
WEPP—Water Erosion Prediction Project
WEPS—Wind Erosion Prediction System
WEQ—Wind Erosion Equation

## EROSIVITY

Erosivity is the ability of rain or wind to cause a soil to erode. Rain erosivity increases with the amount and intensity (total energy) of rain during a given storm. Wind erosivity has been defined as that property of wind that affects its ability to entrain and move soil particles. It generally increases with wind speeds at the soil surface.

## EVAPORATION

Evaporation is the process by which a liquid is changed into a vapor or gas. It accounts for major losses of soil water under certain conditions, thereby constituting a major deterrent to storing water in soil for subsequent use by plants. Evaporation also causes water losses from surface reservoirs. It can indirectly affect erosion by water and wind by making the soil more erosive. For example, excessive water losses through evaporation may result in poor plant growth, thereby making the soil more susceptible to erosion. A dry soil may also be more susceptible to erosion by wind. In contrast, a wet soil where evaporation has been limited may result in greater runoff and, therefore, a greater potential for erosion by water.

## EVAPOTRANSPIRATION

Evapotranspiration is the combining term used to denote the total loss of water from a given area as a result of evaporation from the soil surface and transpiration from plant leaves during the growing season. Similar to evaporation, it may indirectly affect erosion by water and wind (*see* **EVAPORATION**).

# EXPLOITATION OF NATURAL RESOURCES

Exploitation of the world's natural resources has been going on for centuries, usually on lands without major limitations with respect to plant growth. When used wisely, the resources have retained their productivity. With increasing demands to produce food and fiber for an ever-increasing population, exploitation of less desirable lands is leading, or has led, to soil degradation in many places. Such exploitation often results in an increased potential for erosion by water or wind and reduces the potential for water conservation.

# F

## FARM SIZE (FARMER RESOURCES)

Farmers with adequate land and available resources usually can sacrifice some land for installing suitable erosion control practices or managing the land in a manner that helps control erosion. Through such practices, water conservation usually is also enhanced where required. In contrast, farmers with limited land and resources may use all their land to produce the food needed by their families, and taking any land out of production could cause food shortages. Some soil and water conservation practices are applicable to small cropped areas. With proper management, they can be used and still afford farmers the opportunity to produce the food required for their families.

## FENCES

Fences constructed of tree branches, dead shrubs, or manufactured materials (e.g., corrugated asbestos-cement sheets) have been shown to be effective for controlling soil movement caused by wind.

## FIELD LENGTH AND WIDTH

Field size (length and width) strongly influences the potential for soil and water losses and the type of soil and water conservation

© 2006 by The Haworth Press, Inc. All rights reserved.

doi:10.1300/5678_07

practices that can be carried out. In general, the potential for soil and water losses increases with increases in field size. On large fields, almost any type of soil or water conservation practice can be applied. In contrast, a drainage or terrace system that requires an outlet may not be appropriate for use on a small field unless provisions can be made to include the adjacent field (or fields) in the system. This may require cooperation among neighboring landowners (*see* **HUMAN FACTORS**, ***Area Farmed/Owned***).

## FINANCIAL ASPECTS

### *Ability*

The application of some soil and water conservation practices to land involves certain expenses. Hence, the financial ability of the producer or landowner may influence whether appropriate conservation practices are used. Where outside financial assistance is available, as from government programs, it may be possible for the practices to be applied with only a limited financial outlay by the producer or landowner. Such financial assistance is beneficial to society as a whole because the land's productivity must be sustained for future generations to be able to produce the foods, fibers, and so on that they require.

### *Banks*

In some countries, either banks are not available or producers and landowners prefer not to hold their wealth in a monetary form by depositing it in a bank. Instead, their wealth is measured by the number of animals they own. Although this practice is not undesirable in and of itself, it becomes a problem with regard to soil and water conservation when animal populations exceed the carrying capacity of the land (*see* **ANIMALS,** ***Populations***).

### *Credit*

The implementation cost of some conservation practices may require a producer or landowner to seek credit from a lending institution (e.g., a bank). The success of a loan application may determine whether conservation practices are implemented. Unfortunately, some lenders are (or were) reluctant to make normal crop production loans to farmers interested in producing crops using some of the newer conservation practices (e.g., conservation tillage, no-tillage). This denial of production loans could lead to continued soil and water losses.

### *Economic Constraints*

*See* **FINANCIAL ASPECTS,** ***Ability; Credit; Economics.***

### *Economics*

The adoption and application of soil and water conservation practices to land involve economic considerations. Some practices require expenditures, others may reduce expenses, and some may be cost neutral. Practices that involve expenditures include those requiring soil movement by machinery or hand labor (e.g., ditches, canals, waterways, terraces, and embankments) and the construction of facilities to control, minimize, or prevent soil or water movement (e.g., dams, dikes, and buffer strips). A related, indirect expense is incurred when some land is taken out of production to accommodate conservation structures. Practices that reduce expenses are those that can be implemented through changes in management such as soil and water conservation based on retaining crop residues on the soil surface. An expenditure-neutral practice is crop production on the contour rather than up and down the slope of the land. The entire land surface remains in crop production, but the direction of operations across the land surface is altered with little or no change in the cost of operations.

### *Energy Cost*

Energy is required to implement conservation practices. Whether it is fuel for machinery, feed for animals, or food for humans, a cost is involved. Therefore, energy costs compared with the benefits to be obtained are a major factor in how readily appropriate conservation practices are adopted. Satisfactory low-cost practices are available for many conditions, and these should be emphasized where energy costs are of major concern (e.g., in many developing countries).

### *Income*

The level of financial income of a producer or landowner often plays a major role in determining whether soil and water conservation practices will be adopted and applied to land. Although financial assistance from outside sources may be available to implement conservation practices, maintenance of the practices usually becomes the responsibility of the producer or landowner. In addition, the installation of some conservation practices requires that some land be taken out of production. This may result in lower income for the producer or landowner, who may, therefore, choose not to adopt the conservation practices.

### *Markets*

Suitable markets for products from farming and livestock operations are essential for successful production enterprises. With adequate financial returns from the operation, producers and landowners are more likely to have the resources to initiate appropriate soil and/or water conservation practices.

### *Profitability of Adopting Conservation Practices*

Applying soil conservation practices to land often does not result in immediate financial benefits to the producer or landowner, except where erosion causes damage to crops or equipment, for example. However, if erosion remains unchecked, crop productivity usually declines, resulting in reduced profitability. In contrast, improved water conservation may result in immediate benefits because of better crop yields, especially for dryland crops in a semiarid region.

# FOREST

A forest is an extensive tract of land covered by a relatively thick growth of trees and underbrush. It may be managed for timber production, reserved for game animals, or used for recreational purposes. Trees intercept raindrops, dissipating their energy and allowing the water to fall to the ground at a reduced rate. Forest litter further absorbs the energy of the falling water, protecting the land against erosion by water and allowing the water to enter the soil.

## *Afforestation*

Afforestation is the practice of establishing a forest on land by artificial means such as planting or seeding where trees have never grown. Provided enough trees are established, the canopy will dissipate the energy of falling raindrops, reducing the potential for erosion by water and increasing the potential for water conservation. The established trees will also reduce or eliminate erosion by wind.

## *Deforestation*

Deforestation is the removal of trees from forested lands. When deforestation occurs, as for land development for agricultural purposes (crop production), the soil and water conservation benefits of the trees no longer exist, unless appropriate conservation measures are immediately applied. Deforestation may also occur to some extent when trees are harvested for lumber using the clear-cutting method, where all trees are removed from a given area. This may result in erosion and water losses until newly planted trees again protect the land. Litter and understory vegetation left on the clear-cut area may also provide some protection against erosion, even before the newly planted trees do.

## *Litter*

Forest litter is the accumulation on the soil surface in a forest of unincorporated tree leaves, twigs, and the like. Litter intercepts the energy of falling raindrops, which minimizes the dispersion of surface soil aggregates and retards water flow across the surface. This provides more time for infiltration. As a result, the potential for runoff that could result in erosion by flowing water is reduced and the potential for storing water in soil is enhanced.

### *Reforestation*

Reforestation is the practice of reestablishing trees on forest land from which trees have been removed for lumber or destroyed by fire or insects, for example. Reforestation reduces the potential for soil and water losses.

## FREEZE-THAW CYCLE

Freezing and then thawing of a wet soil usually results in the dispersion of surface soil aggregates and often leaves the surface soil in a "fluffy" condition. When aggregates are dispersed, a surface seal quickly forms when rain occurs. This can result in additional runoff that increases the potential for erosion by water and decreases the potential for water conservation. A fluffy soil often is highly susceptible to erosion by wind.

# G

## GENETIC ENGINEERING

Genetic engineering is the practice of altering the genetic makeup of an organism to introduce a trait different from one originally inherent in the organism. The development of herbicide-resistant crops allows herbicides to be applied to control weeds in an established crop, which reduces or eliminates the need for tillage for weed control. Under some conditions, the potential for erosion by water and/or wind is reduced by using herbicides rather than tillage for controlling weeds. Eliminating tillage also reduces the potential for soil water evaporation, which, along with reduced water use by weeds, increases the potential for greater soil water conservation.

## GENETICS (IMPROVED CROPS)

Whether through natural selection, cross-breeding, hybrid development, or genetic engineering, crop improvement can affect soil and water conservation in a number of ways. These ways include improved adaptability of the crop to prevailing conditions, which results in more rapid development and better surface cover to protect the land against erosion; greater supply of crop residues after harvest, which can be managed for soil and water conservation purposes; and greater monetary returns to the producer or landowner, which provides the incentive to apply conservation practices to the land.

© 2006 by The Haworth Press, Inc. All rights reserved.

doi:10.1300/5678_08

# GOVERNMENTAL FACTORS

Various local, district, state or province, and national governmental factors directly or indirectly influence the availability, adaptability, and implementation of appropriate practices for conserving soil and water on a given tract of land.

## *Aspirations*

The aspirations of government leaders (elected or appointed officials) may strongly influence the emphasis (or lack of it) given to soil and water conservation in a given country, state or province, or local area. Leaders who recognize the importance of preserving natural resources for long-term crop productivity and food security generally will strive to conserve soil and water resources. In contrast, those who place strong emphasis on industrial enterprises may rely on food importation and neglect resource conservation.

## *Assistance—Technical/Practical/Research*

Most producers or landowners rely on assistance from some government agency in selecting and implementing appropriate conservation practices to a given tract of land. The assistance may include technical help and practical advice to select the most efficient and effective practice and then to properly construct or install, operate, and maintain it. The selected practices should have been evaluated under appropriate research conditions; often this research is financed by a government agency.

## *Conflicting Programs*

The agricultural programs of a government may support soil and water conservation, even to the extent that funding is allocated to implement conservation practices. When conservation practices are applied to land, it may be necessary to take some land out of crop production (e.g., for terraces or waterways). Later, the government may be interested in greater crop production and encourage the use of all available land for that purpose. Producers may forgo using conservation practices, especially if crop prices are favorable. Such situation occurred in the United States in the 1970s, when farmers were encouraged to grow crops "fence row to fence row." Similarly, increased crop production is encouraged during wartime, as was the case in the United States during World Wars I and II.

### *Crop Regulations*

The types of crops grown, the amount of land devoted to a specific crop, and when a crop must be destroyed (e.g., to control crop pests such as insects and diseases) may be mandated by governmental regulations. Unless their consequences are clearly understood, such regulations may adversely affect soil or water conservation efforts.

### *Export-Import Balance, National*

The export-import balance on a national basis may indirectly influence the potential for erosion by water or wind. When export demand for a given crop product is strong, production is encouraged, which may result in marginally or even highly erosive lands being used to produce that crop. As a result, the potential for erosion is increased. With greater imports, marginal or highly erosive lands may be taken out of production, thereby reducing the potential for erosion.

### *Farm Programs, Federal*

Federal farm programs may have positive or negative effects on soil and water conservation. In the United States, the Conservation Reserve Program established as part of the Food Security Act in 1985 and the Great Plains Conservation Program established in 1956 were designed to take highly erodible land out of crop production and to manage the land to control erosion by planting grasses or trees. Landowners were compensated for their participation in the programs. In contrast, some farm programs encourage production, which may lead to crop areas being expanded, often onto highly erodible land where the potential for erosion may be increased if inappropriate farming practices are used (*see* **GOVERNMENTAL FACTORS**, ***Conflicting Programs; National Policies; Politics***).

### *Legislation and Ability to Enforce Laws*

Many land use activities related to soil and water conservation are regulated by laws passed by national, state or province, or even local legislative entities. For the regulations to be effective, the legislating entity must have the ability to enforce them, which sometimes is not the case because of personnel or financial limitations. As a result, the level of conservation visualized for the legislation may not be achieved.

### *National Policies*

National policies can strongly influence the degree to which soil and water conservation practices are selected, implemented, and maintained in a country. For example, if the emphasis is on maximizing food production by using virtually all available land to grow food crops, then marginal, highly erodible land may not be adequately protected against erosion. In contrast, if food production is adequate, such erodible land may be taken out of production and used to grow plants that help control erosion.

### *Political Constraints*

*See* **GOVERNMENTAL FACTORS, *National Policies; Politics***.

### *Politics*

The national policies that affect soil and water conservation are, to some extent, influenced by the dominant political party in a country. However, although the party in control may advocate resource conservation, projects such as road construction and education usually have a more immediate impact on voters and, therefore, are more likely to be funded than conservation projects.

### *Subsidized Conservation Works*

A subsidy is a government grant to a private enterprise considered to be of benefit to the public. With regard to soil and water conservation, a subsidy is a payment to an individual or cooperative farming enterprise for implementing (and possibly maintaining) conservation practices. The justification for the subsidy is that the general public also benefits from conserving soil and water. The payment may not cover the full cost of implementing the practices. In many cases, a government agency provides technical implementation assistance, which amounts to an indirect subsidy from the government. Subsidies have been highly effective in encouraging the use of soil and water conservation practices in many cases.

### *Taxes*

Taxes are the main source of revenue used by government entities in most countries to fund projects and activities deemed to be beneficial to the country as a whole, including soil and water conservation. As agricultural producers and landowners generally are not exempt from paying, taxes may have an indirect effect on soil and water conservation. Tax relief is granted in some countries when certain conservation practices are implemented, which may provide an incentive for producers and landowners to implement such practices.

## GRASS COVER IN PLANTED CROPS

Maintaining a grass cover in a growing crop (e.g., Bermuda grass in corn) helps conserve soil and water by protecting the soil surface, intercepting raindrops, and reducing the rate and amount of runoff. One disadvantage of such grass cover is that the grass uses plant nutrients and soil water, which may under some conditions deprive the crop of the adequate supply of nutrients and water needed for an optimum yield.

# HERBICIDE-RESISTANT CROPS

*See* **GENETIC ENGINEERING.**

# HUMAN FACTORS

A number of human factors have a direct or indirect effect on soil and/or water losses and conservation.

### *Age of Producer/Landowner*

The age effect results mainly from the education and attitudes of the person. In general, younger people often are better educated, aware of the practices and policies that affect soil and water conservation, and willing to use them (take risks) to achieve the desired results than older people. In contrast, older people with more limited education may be reluctant to change how they manage their farms to achieve improved soil and water conservation.

### *Area Farmed/Owned*

Erosion and water conservation can be relevant on any farmed area, from small tracts farmed using hand implements in some countries to farms covering thousands of hectares farmed with massive machines in other countries. Some practices to control erosion by wa-

© 2006 by The Haworth Press, Inc. All rights reserved.
doi:10.1300/5678_09

ter and for water conservation (e.g., terraces, contour tillage, and use of waterways), however, may not be suitable for small tracts of land unless neighboring producers cooperate in their implementation. In contrast to erosion by water, the susceptibility of land to erosion by wind generally increases with the size of the tract being farmed. Small tracts, therefore, may not be subject to erosion by wind if their boundaries are stable and provide resistance to wind flow at the soil surface. Large tracts under similar soil and climatic conditions, however, may be highly susceptible to erosion by wind.

### *Attitude*

The attitudes of producers, landowners, and society as a whole strongly influence whether soil and water conservation practices will be used. Some producers and landowners recognize the effect of erosion on their land and its effect on their livelihood. Likewise, where water conservation is important, they realize the value of conserved water for improving crop yields. As a result, they evaluate the costs and benefits of implementing conservation practices and act accordingly, from the point of view of both direct costs and resource conservation. Such producers and landowners often readily adopt appropriate practices, make adjustments in their management practices to accommodate these practices on the land, and possibly relinquish some production if land is needed for installing the conservation practices. In contrast, some producers and landowners strive for maximum production without concern for potential soil and water losses. Although immediate production may not be greatly affected by the losses, the long-term productivity of the land usually declines following uncontrolled erosion. In other cases, people not directly connected with the farming enterprise do not act because they view the erosion problem (e.g., gullies; sediment deposition on roads, in streams, and in reservoirs; dust storms) as a problem of the producers and landowners. However, the problem is costly for them also because of greater taxes for cleanup, damage to roads and equipment, accidents, and, in some cases, health problems. Others view the erosion problem as the responsibility of society as a whole; they realize that uncontrolled erosion has costs and leads to eventual loss of soil resources, which may result in food shortages and increased food prices. As a result, they seek and support programs that provide assistance in controlling erosion. Although there is a cost involved in such programs, they realize that controlling the problem will benefit not only the producers and landowners but society as a whole. The attitudes, goals, and actions of groups within a society at the local, district, state or province, and national levels often have a major impact on soil and water conservation in a given area, region, or country. Such groups include local conservation districts, producer organizations, and professional societies. In addition, private citizens sometimes identify problem areas lacking effective soil and water conservation and unite as a group to express their concerns and work with appropriate agencies to implement measures to overcome the problems.

### *Encroachment*

Encroachment results from population pressures and the need for land; it involves movement onto lands originally retained, often by a country's leaders, for other purposes. Such land often was reserved because it was not ecologically suitable for crop production (e.g., steep slopes). Encroachment often leads to severe erosion by water and poor water conservation because appropriate conservation practices seldom are used under such conditions.

### *Health/Nutrition*

Indirectly, the health and level of nutrition of producers and landowners and their families can affect their interest in soil and water conservation and their ability to make the financial commitment to adopt, apply, and maintain appropriate conservation practices. Those enjoying good health and nutrition are more likely to have the financial resources required. Those with poor health and nutrition may require much or all of their available resources to maintain their health and provide food for themselves and their families.

### *Labor*

In some countries, certain soil and water conservation practices involve a considerable amount of labor. Therefore, labor availability can have a major impact on whether appropriate and adequate conservation practices are used to conserve soil and water.

### *Lack of an Appropriate Way to Initiate Action*

Soil and water conservation practices range from simple to highly complex. Although producers and landowners may realize the need for conservation practices, they may not have the knowledge and/or equipment required to install even a relatively simple practice successfully, much less a complex one. Therefore, unless technical assistance along with access to appropriate equipment is provided by some outside source, conservation practices may not be implemented.

### *Lack of Awareness of a Problem*

Slight soil and water losses sometimes are not easily detected. Even when they are, some producers and landowners do not consider such losses to be a problem and, therefore, do not adopt appropriate conservation practices. Continued small-scale losses can, however, eventually result in severe soil degradation.

### *Lack of Trained Staff*

Although the need for conservation practices may be realized, such practices may not be implemented because a staff of knowledgeable people needed for successful implementation of appropriate practices for the given conditions may not be available.

### *Nomads/Migration*

In some countries, nomads migrate across the countryside in search of forage and water for their grazing animals. Their animals are allowed to graze or forage on fields after the planted crop has been harvested. This frequently results in little or no crop residue being retained on the soil surface, especially when the animals are sheep or goats. As a result, the land may become highly susceptible to erosion by water and wind and have a low potential for water conservation.

### *Population Pressure*

Growth in human populations is resulting in increased potential for soil and water losses because of increased cropping intensity on land already being cropped and the expansion of cropping to lands less suitable for crop production or more susceptible to erosion. More intensive cropping results in the same land being planted to crops more frequently, often without any of the land being set aside for natural vegetation or a cover crop to protect the soil or to replenish its nutrient supply (e.g., shorter noncrop periods where shifting cultivation is practiced; *see* **CROPPING SYSTEM/SEQUENCE**, ***Shifting Cultivation***).

### *Poverty*

Interest in and funding for soil and water conservation usually is low or even nonexistent where people live in poverty. The primary concern is for food and shelter. Any conservation under such conditions would most likely be done by some outside entity (e.g., government or social agency).

### *Short-Term Motive*

The short-term motive (or goal) of an individual (e.g., producer, landowner), organization, or government may influence the potential for soil and water conservation. For example, if maximum production of a crop is desired or encouraged, it may be grown on land and under conditions not conducive to conserving soil and water. Of course, the short-term motive also may be conducive to conservation, as would be the case, for example, where special efforts are made to control soil and water losses at a construction site.

### *Social Constraints*

Social constraints that may have a bearing on soil and water conservation include such factors as land ownership (individual, communal, or government), land tenancy (short versus long term), and lifestyle (settled, nomadic, or shifting cultivation). Although there are exceptions, soil and water conservation is more likely to be practiced by people who own the land, have long-term tenancy rights, or have a settled lifestyle.

### *Social Effects—General*

Social influences on crop production systems and associated land management practices are numerous and varied and may result in soil and water losses in some cases and conservation in others. The social factor most responsible for erosion is probably the increasing world population. Others include ownership of more animals (e.g., cattle, sheep, goats) than can be easily fed with existing forages, tolerance of or failure to control animals that do not contribute to the food supply, emphasis on the production of land-degrading crops, and the introduction of practices that are not suitable under the prevailing conditions. Soil and water conservation can be achieved when society recognizes its value and provides incentives for the adoption of appropriate conservation practices or else imposes penalties for non-adoption of such practices.

### *Willingness of Producers*

Achieving soil and water conservation requires the dedication of all those with an interest in a given tract of land to "getting the job done." When a new conservation practice is to be used, it may be necessary to modify or even abandon crop production practices that have been used for a long time. This requires that the person on the land (the producer) must be willing to make the necessary operational changes so that the new conservation practices will be correctly applied to achieve the desired results.

# I

## INFILTRABILITY

Infiltrability is the flux (or rate) at which water moves downward into a soil. To determine the infiltrability of a soil, water at atmospheric pressure is maintained at the atmosphere-soil boundary and its rate of movement into the soil is measured. High infiltrability is conducive to reduced runoff, which reduces the potential for erosion by water and increases the potential for water conservation.

## INFILTRATION

Infiltration is the process of water entering a soil and has a direct influence on erosion by water and water storage in soil. Water that infiltrates a soil does not flow across the surface, where it may cause erosion. Wind erosion may be indirectly affected: greater infiltration (improved water conservation) can lead to improved conditions for plant growth and more vegetation on the surface to protect soil against the forces of wind.

## INFORMATION

Reliable information forms the basis for successful soil and water conservation. The following factors are important for implementing, achieving, and maintaining successful conservation practices.

© 2006 by The Haworth Press, Inc. All rights reserved.
doi:10.1300/5678_10

### *Access to Information Systems*

Much information has been developed that can be used to help control erosion and conserve water. Most of it is publicly available because it was developed by public agencies. In developed countries, such information usually is available in libraries, and help regarding the implementation of soil and water conservation practices can be obtained from public agencies such as the Natural Resources Conservation Service (federal) and Extension Service (state) in the United States and similar agencies in other countries. Access to such information may be limited in less developed countries, especially if the level of education is low, little or no emphasis is placed on soil and water conservation, public agency personnel are not adequately trained, and financial support for conservation efforts is lacking.

### *Education/Extension/Training*

For effective conservation of soil and water, appropriate education is essential at all levels of society (from people on the land to those responsible for designing, constructing, and financially supporting the implementation of conservation practices). People on the land must recognize the need for conservation practices or be made aware of this need through extension activities or appropriate training. Likewise, technical personnel must have the appropriate education to recognize what is needed, design appropriate conservation practices, and ensure that these are properly implemented and maintained. Those responsible for supporting conservation efforts must understand the need for them, the costs involved, and the means of obtaining financial support for implementation. The level of education of the producer or landowner may or may not influence which conservation practices are applied to the land.

### *Research*

Most current soil and water conservation practices are the result of research at some location. Although some practices were designed for a particular location or problem, with relatively minor modifications they often can be extended to other locations through careful evaluation of the problems. Although sound research is fundamental to developing effective conservation measures, it is also true that some effective practices have been developed by people who are not researchers but who recognized a problem and took actions to control it.

### *Technology*

As used in this book, "technology" is the science or study of practices for controlling erosion by water and wind and for conserving water. Most practices that are currently recommended and used are based on technology. Some older practices (some of which are still in use), however, were based on producer or landowner ingenuity and provided for good erosion control and water conservation.

### *Training Programs*

*See* **INFORMATION,** ***Education/Extension/Training.***

## INFRASTRUCTURE

Infrastructure is the term used to denote the basic installations and facilities on which a community, district, state or province, or nation depends for continuance and growth. Infrastructure includes roads, schools, power plants and transmission lines, communication lines, and oil and gas pipelines. At some time, the development of each of these types of facilities required disturbance of the natural landscape, which may have resulted in the potential for erosion. Even after satisfactory installation, the potential for erosion may still exist at some facilities. As a result, continued careful management is required to minimize or avoid the occurrence of damaging erosion.

# IRRIGATION

Irrigation is the application of water to a soil to enhance crop productivity. The method of application and the conditions under which water is applied may have a direct effect on erosion by water and the efficiency of water application, with the latter affecting water conservation. With improved plant growth, erosion by water generally is reduced: greater cover reduces raindrop impact on the surface, thereby reducing surface sealing, improving infiltration, and retarding runoff across the surface. Irrigation may also increase erosion by water, depending on the rate and method of water application (*see* **EROSION, *Irrigation Erosion***). Erosion caused by irrigation water may be increased when rain occurs shortly after irrigation. Irrigation's effect on erosion by wind is through its effect on plant growth and possible carryover of crop residues, which can be managed to control erosion by wind during the interval between crops. Without irrigation, a soil may be highly susceptible to erosion by wind. The following irrigation methods have direct or indirect effects on soil and/or water conservation.

## *Alternate Furrow Irrigation*

Alternate furrow irrigation is the practice of applying water to alternate furrows for crops planted in rows. This results in lateral movement of water to plants, often without wetting the entire soil surface, thereby reducing water losses through evaporation. In some cases, the furrows receiving water are switched at successive irrigations.

## *Basin Irrigation*

Basin irrigation is the practice of applying water to relatively level areas surrounded by small earthen levees or dikes. It is a type of flood irrigation and is best suited to soils with a relatively low infiltrability (*see* **IRRIGATION, *Flood Irrigation***).

### *Corrugation Irrigation*

Corrugation irrigation is the practice of applying water to small, closely spaced furrows, called corrugates, which cause water to flow in one direction. Corrugation irrigation is frequently used for grain and forage crops.

### *Deficit Irrigation*

With deficit irrigation, crop plants are allowed to develop a certain level of water stress either during particular plant growth stages or throughout the entire growing season. Water use is reduced, water use efficiency generally is increased, and the reduction in crop yields is relatively small compared with the benefits gained by saving water, which can be applied to other crops.

### *Drip Irrigation*

Drip irrigation involves the slow application of water to soil through emitters with low-discharge orifices. The water can be applied on the soil surface, which reduces the potential for runoff and erosion by water, or below the surface, which greatly reduces the potential for evaporative losses and enhances water conservation.

### *Flood Irrigation*

For flood irrigation, water is released from ditches or pipes and allowed to flow over (flood) the entire tract of land being irrigated. Flood irrigation is best suited for land with a low surface slope and soil with a relatively low water infiltration rate. Surface leveling is sometimes carried out to improve water distribution over the land area.

### *Full Irrigation*

Full irrigation is the practice of applying an amount of water to a crop equal to that removed from the field by evapotranspiration (transpiration by the crop plus evaporation from the soil surface). Although full irrigation generally results in the highest crop yields, water use efficiency may be reduced and the potential for erosion may be increased if precipitation occurs soon after an irrigation.

### *Furrow Irrigation*

In contrast to alternate furrow irrigation, furrow irrigation involves water application to all furrows at each irrigation. Water is applied at the upslope end of the field and allowed to flow downslope through the field. Furrow irrigation is suitable mainly for land with a relatively low surface slope and soil with a relatively low water infiltration rate.

#### *Broad Furrow Irrigation*

For broad furrow irrigation, the furrow spacing is greater than that normally used for a given crop (e.g., winter wheat).

#### *Contour Furrow Irrigation*

For contour furrow irrigation, furrows are oriented across the slope of the land. The furrows have a slight downward slope gradient to cause water to flow through the field.

#### *Graded-Furrow Irrigation*

For graded-furrow irrigation, furrows are laid out on land that has been altered (graded) to provide a downward slope that provides for more uniform water distribution throughout the field.

### *Limited Irrigation*

Limited irrigation is the practice of applying less water than is needed to replenish the water used by the crop since the previous irrigation. Although this practice generally reduces crop yields, the efficiency of water use by the crop often is greater with limited than with full irrigation (*see* **IRRIGATION, *Full Irrigation***).

### *Management*

Many factors must be considered for efficient and effective use of irrigation water for crop production. With appropriate management, excellent results in terms of crop production and soil and water conservation are possible. With poor management, highly unfavorable results can occur.

#### *Land Preparation*

In many regions and for certain methods of applying water, some land preparation is required, for example, the installation of a water delivery system, installation of levees or dikes, leveling or grading of the land surface, and furrowing of the land surface. Each can have an impact on soil and water conservation.

#### *Land Slope*

The slope of land strongly influences which method of irrigation can be used at a given site. Careful evaluation is needed to select the most effective system for the prevailing conditions to minimize the potential for soil and water losses. Furrow irrigation generally is suitable for slightly sloping land, whereas sprinkler irrigation is suitable for a wide range of surface slopes (*see* **IRRIGATION, *Furrow Irrigation; Sprinkler Irrigation***).

#### *Scheduling*

Irrigation scheduling is critical for the most efficient use of water for a given crop. Different crops require water at different intervals, with irrigations being more important at some growth stages than at others for achieving optimum yields. Scheduling may also be influenced by water availability, namely, when water is allocated from some outside source (e.g., a reservoir) by a water district or other agency.

*System Design*

Most methods of applying water have been developed for a particular situation and are not universally applicable to all situations. Careful management, therefore, is necessary to select the most appropriate method for a given situation to achieve the greatest benefits in terms of irrigating the crop and conserving soil and water.

*Water Delivery System*

The water delivery system used has a major effect on water conservation and may affect erosion by water. Water can be delivered from the source (reservoir, stream, well) to the point of use by a canal (lined or unlined), open ditch, aboveground pipe, or underground pipe. Each method is satisfactory under some conditions but may be totally inappropriate under other conditions. Careful management is needed to evaluate the conditions fully and select the best method for conveying the water.

### ***Sprinkler Irrigation***

For sprinkler irrigation, water is applied over the entire soil surface through spray nozzles or high-volume guns using a pressurized system. Sprinkler systems are available that are suitable for a wide range of land surface slopes. Methods of water application using sprinkler systems include high pressure, which applies water to a relatively large area but may result in relatively high evaporative losses under some conditions, and low pressure, which uses drop nozzles or other means of applying water close to the surface over a relatively small area. The LEPA (low-energy precision application) method results in water application efficiencies greater than 90 percent, which has a major water conservation benefit. Types of sprinkler systems include:

*Boom System*

For a boom system, an elevated, cantilevered sprinkler (or sprinklers) is mounted on a central stand on which the sprinkler boom rotates as water is applied. The traveling gun is a boom type that is mounted on a frame and delivers water supplied though a flexible hose. The gun is manually or automatically pulled by a cable through the field.

*Center Pivot System*

For the center pivot system, a pipe supplies water to sprinkler nozzles or heads at appropriate spacings on the pipe, which is supported above the crop by towers at fixed distances. The pipe supported by towers automatically rotates around the center or pivot point at which water is supplied to the system. The towers are mounted on wheels or skids and are propelled by hydraulic, mechanical, electric, or pneumatic power.

*Lateral (or Linear) Move System*

The components of a lateral move system are similar to those of the center pivot system, except that the system moves laterally with water supplied through a drag hose attached to one end of the pipe. Variations of lateral move sprinkler systems can be found:

1. For the side-move type, the supply pipe is supported on carriages to which trailing small-diameter pipelines fitted with several sprinkler heads are attached.
2. For the side-roll type, the supply pipe is mounted on wheels, with the pipe serving as the axle for the system.

*Permanent System*

For a permanent irrigation system, water is applied through risers attached to permanently installed underground pipes.

*Solid Set System*

A solid set system covers the entire area, allowing repeated irrigations of the entire area through surface pipe and sprinklers without moving any of the system.

### *Subirrigation*

With subirrigation, water is applied to the bottom of plants and the plants absorb the water. This method is common for irrigating bedded plants in greenhouses and for landscaped plants such as those growing in containers. Subirrigation can also be achieved by raising the water level in field areas. Because the water is applied below the surface, the potential for erosion by water is minimal. Drip irrigation applied below the soil surface is a type of subirrigation (*see* **IRRIGATION, *Drip Irrigation***).

### *Supplemental Irrigation*

Supplemental irrigation is any irrigation that is used to achieve satisfactory plant establishment or to enhance the productivity of crops for which the primary water source is precipitation or water previously stored in the soil.

### *Surge Irrigation*

Surge irrigation is the practice of applying water intermittently during a single irrigation event. Water is applied to one part of a field at such a rate that it flows quickly toward the lower end of the field; it is then applied to the other part of the field after a predetermined time interval. Infiltration occurs before water is applied again to each part.

### *Trickle Irrigation*

*See* **IRRIGATION, *Drip Irrigation*.**

# L

## LAND

Land is the entire complex of surface and near-surface attributes of the solid portions of the surface of the earth. In some classification systems, land includes the bodies of water located within land masses. Land is where erosion by water and wind occurs and where water conservation is important in many places. Many attributes of land and activities involving the land strongly affect the potential for erosion and the need for water conservation.

### *Abandoned Cropland*

Whether abandoned cropland is subject to erosion by water or wind initially depends on the nature of the plant cover when the land is abandoned. If adequate crop materials such as straw, stover, or other residues remain on the soil surface, the land probably is adequately protected against erosion, which then would also provide water conservation benefits because runoff is retarded. Subsequent erosion can be controlled or avoided if the land is planted to grass. If few or no crop materials remain on the surface and grass is not planted, erosion by water or wind may be serious initially, and water conservation will be hindered. With time, weeds and other plants may become established, which would then tend to minimize erosion, unless gullies resulting from erosion by water and blowout areas resulting from erosion by wind develop before a vegetative cover becomes established.

© 2006 by The Haworth Press, Inc. All rights reserved.
doi:10.1300/5678_11

### Allocation

Land allocation is the process of deliberately providing land to nonlandowners. In some countries, this has been achieved by subdividing some previous large landholdings into small units, either after purchase from previous owners or by government policies.

### Alluvial Fan/Plain

Alluvial fans or plains are a consequence of erosion by water. Fans or plains consist of sediments transported by water from upstream (or upslope) points where erosion occurred and deposited at downstream (or downslope) points where the water flow rate was no longer adequate to transport the sediments. When sediments are deposited on cropland, crop productivity often is reduced. Such sediment deposition may also occur on roads and highways, resulting in hazardous travel conditions and requiring cleanup and maintenance expenditures.

### Capability

Land capability is the suitability of land for different uses without permanent damage. In the United States, it is an expression of the effect of the physical features of land, including climate, on its suitability for use for crops, grazing, woodland, or wildlife without causing damage. The risk of damage by erosion and other causes owing to an area's physical characteristics is used in assessing land capabilities.

### Capability Classes

In the United States, eight land capability classes are recognized, with four classes being suitable for cultivation (Class I has few limitations and Class IV has very severe limitations). The remaining classes (V through VIII) are not suitable for cultivation without major treatments. Class V land has little or no erosion hazard but has other limitations that make it unsuitable for cultivation. Limitations increase for the remaining classes, with Class VIII restricted to recreation, wildlife, water supply, or aesthetic uses.

### *Change in Ownership*

A change in land ownership may have positive or negative effects on soil and water conservation, depending on such factors as the attitude, age, education level, and preferences of the new owner compared with those of the original owner (*see also* **LAND, *Change in Use***).

### *Change in Use*

Well-managed grass and forest lands generally are the least erodible landscapes in a given region. When the use of such land is changed (e.g., converting them to cropland), it often becomes more erodible unless appropriate conservation measures are implemented as the land is converted. When erosion occurs, water conservation usually is decreased. Other changes that frequently lead to severe erosion problems, at least until the projects are completed, include land developments for residential, commercial, and industrial uses; road and highway construction; tree harvesting operations; and mining operations. Some success in controlling erosion under such conditions can be achieved by using appropriate ground covers such as mulches or planting suitable grasses (*see* **MULCH/MULCHING**). For maximum effectiveness, such practices should be implemented immediately after (or in some cases before) changes in land use are started.

### *Classification*

Land classification is the process of assigning land units to various categories based on an assessment of the properties of the land and its suitability for particular purposes. An important consideration in the process should be the potential effect of using the land for a particular purpose on soil and water conservation.

### *Consolidation of Landholdings*

In some countries, dividing land when ownership passes on to heirs can result in an heir receiving several nonadjacent tracts of the original land (*see* **LAND, *Fragmentation***). Applying conservation practices to any small tract of land is difficult (*see* **HUMAN FACTORS, *Area Farmed/Owned***); it becomes even more difficult when several nonadjacent tracts are involved. To reduce the problem, land can be traded or purchased to consolidate a person's landholdings, which should make application of conservation practices easier.

### *Critical Areas/Slopes*

Some land areas are more prone to erosion by water or wind than other areas. The potential for erosion by water generally increases as land slopes increase, and some soils may be more erodible than other soils. As a result, a thorough knowledge of the landscape is important for determining which practices are needed for controlling erosion on a given tract of land. Likewise, the need for conserving water may vary across the land, with water conservation usually being more difficult on sloping than on relatively flat land.

### *Desert Pavement*

Desert pavement is the land surface condition that results where erosion by wind (also possibly by water) has removed virtually all soil materials, thereby leaving the soil surface covered with gravelly or stony materials. Once the desert pavement condition has developed, further erosion, especially by wind, is greatly reduced. Erosion by water should also be reduced, unless concentrated flow dislodges the pavement during intense rainstorms. If the pavement reduces runoff, then water conservation should occur also.

### *Erosion Pavement*

*See* **LAND, *Desert Pavement*.**

### *Flood Plain*

A flood plain is the nearly level land adjacent to a stream that becomes inundated during a flood unless protected by some structure such as a levee. The soil of a flood plain consists of sediments deposited by previous floods and is subject to erosion when water flows through the flood plain.

### *Fragmentation*

Land fragmentation occurs when a tract of land is divided, as when ownership is passed on to heirs and their heirs (*see* **LAND, *Consolidation of Landholdings***). Some land sales may also cause land fragmentation. Intensive land fragmentation generally increases the difficulty of initiating soil and water conservation practices.

### *Grading*

*See* **LAND, *Smoothing.***

### *Idle Land*

Idle land is land that has been taken out of production for some reason. Examples are land not needed to meet the current food production requirements in a country, highly erodible land, and land that cannot be managed economically for productive purposes. Such land, if it is not properly managed, can be highly erodible by water and wind and highly ineffective for conserving water. With proper management (e.g., by establishing grass), erosion by water and wind can be almost totally eliminated, as has been the case for land idled under the terms of the Conservation Reserve Program established in the United States in 1985.

### *Imprinting*

*See* **SOIL, *Imprinting/Patterned Soil.***

### *Landform*

A landform is a natural, discernible feature of the landscape such as a mountain, plain, plateau, or valley that exists because of some geological activity. Landforms differ in their potential for erosion by water and wind and for water conservation.

### *Landforming*

Landforming is the practice of shaping the surface of land with tillage or other tools to create a desired surface configuration. Practices range from the use of tillage to form ridges, contours, or furrows in preparation for crops on a small scale to the use of heavy equipment to fill gullies, construct terraces, and build highways and roads on a large scale.

### *Landholding*

A person's landholding is the land which that person owns (*see* **HUMAN FACTORS,** ***Area Farmed/Owned***).

### *Landscape*

Landscape is a term used to denote the features of the land surface that the eye can comprehend in a single view. Types of landscapes differ in their potential for soil and water losses and, therefore, in the potential need to initiate and maintain soil and water conservation practices on them.

### *Leveling*

Land leveling is the process of shaping the land surface for such purposes as improved water distribution and more efficient machinery operation. Marginal lands often require drastic measures to increase their productivity.

### *Marginal Lands*

Marginal lands are those lands at or near the lower limit of their suitability for a particular use. Plant growth on marginal croplands often is poor. As a result, the potential for erosion by water or wind may be high and the potential for water conservation may be low.

### *Meadow*

A meadow is a tract of land devoted to the production of forage (grasses or legumes) either as hay or for grazing by animals. The land may be used as a meadow for a relatively short time, as in a rotation with other crops, or permanently. In either case, the potential for erosion is low because the grasses or legumes adequately cover the soil surface, which retards runoff and shields the soil from the forces of winds. As runoff is retarded, the potential for water conservation is increased.

### *Mound*

A mound is a natural elevation (small hill) or artificial heap or bank of earth. Because of its relatively steep sides relative to the surrounding land surface, the soil of a mound may be more susceptible to erosion by water and may also have a relatively low potential for water conservation. Depending on the soil and vegetative cover, erosion by wind may also occur on a mound.

### *Overwash*

Overwash is material that has been eroded and carried by water from upslope sites or from upstream areas of a region and deposited on the land. The overwash material may or may not improve the productivity of the soil where the material is deposited.

### *Owner-Tenant Relations*

Agreement between the landowner and tenant is essential for implementing and successfully maintaining soil and water conservation practices. In this regard, factors such as the attitudes of both parties toward conservation practices, cooperation in getting the practices implemented, and length of lease are important.

### *Ownership*

Direct ownership does not ensure that a given tract of land will be fully protected against soil and water losses, but it generally results in better protection than when communal ownership or various forms of land tenure are involved. Under communal ownership conditions, mismanagement often occurs that leads to overgrazing by animals or to excessive use of wooded areas for firewood, thus leading to land degradation. When tenants have only a short-term cultivation right or use the land as sharecroppers, their interest in initiating and maintaining long-term conservation projects may be limited.

### *Past Neglect in Management*

What has previously occurred on a tract of land has a direct effect on its condition at the present time. If conservation was neglected and erosion occurred because of improper management in the past, some remedial actions will be required to restore the land's productivity.

### *Reclamation*

Land reclamation is the process of changing the land's general character to make it suitable for more intensive use. Processes include draining wetlands, irrigating arid and semiarid lands, and recovering submerged land from seas and lakes. In all cases, the actions taken should be conducive to conserving soil and water, as required.

### *Reconstruction*

Land reconstruction is the process of altering the surface shape of mined lands to increase their potential for beneficial uses and decrease the potential for soil and water losses.

### *Rehabilitation*

Land rehabilitation refers to the restoration of land degraded by, for example, various forms of agriculture, overexploitative logging, forest and grassland fires, slash-and-burn agriculture, surface mining operations, and military operations (wars and training exercises). The aim of rehabilitation is to prevent further land degradation in environmentally sensitive areas, restore the land to productive uses, and reduce poverty. Rehabilitation reduces the potential for further land degradation and, under some conditions, improves trafficability on the land by eliminating gullies resulting from previous erosion.

### Rejuvenation

Rejuvenation is the process of restoring to a more productive condition a tract of land that may have, for example, sparse vegetation, high runoff, low infiltration, and high rates of erosion by water and wind. To achieve satisfactory rejuvenation, the problem(s) causing the poor condition must be identified and appropriate changes must then be made. Possible corrective actions include altering the surface to decrease runoff (e.g., furrowing, range pitting, or mulching), adding fertilizer, removing a chemical imbalance that inhibits plant growth, or possibly establishing an improved cultivar of the existing plant species or a different species better adapted to prevailing conditions.

### Restoration

Restoration is the process of restoring a site to the conditions that existed before it was disturbed, thereby improving its potential for greater production. Depending on the site, restoration may involve shaping to remedy nonuniform surface conditions; ripping, subsoiling, or disking to improve water infiltration; and fertilization. Through successful restoration, a site becomes more productive and soil and water conservation usually is improved.

### Slope Steepness

Slope steepness strongly influences the potential for erosion by water and for water conservation on a given tract of land. On gentle slopes, practices such as contour tillage and furrow diking may be satisfactory. On steeper slopes, practices such as terracing, strip cropping, and retaining crop residues on the surface may be necessary. To control erosion under steeply sloping conditions, it may be necessary to use the land for forage or timber production.

### *Smoothing*

Land smoothing is the process of moving soil from high to low places in a field to improve the uniformity of water distribution and storage in soil, reduce the potential for erosion, especially by water, and improve the performance of cultural operations.

### *Subdivision*

*See* **LAND,** ***Fragmentation***.

### *Tenure Relations*

*See* **LAND,** ***Owner-Tenant Relations***.

### *Topography*

The topography of an area refers to the relative positions and elevations of natural and built features that describe the configuration of the surface. Surface slope is a feature of topography that has a profound effect on the potential for erosion by water and for water conservation in a given area. Slope also strongly influences the soil and water conservation practices that are suitable for a given tract of land. A variety of conservation practices generally are suitable where surface slopes are relatively low and uniform, but options are greatly limited where, for example, steep and irregular slopes, rock outcrops, broken terrain, or sinkholes are present.

### *Use*

Land use refers to all aspects of how a given tract of private or public land is used (*see* **LAND,** ***Use Planning/Policies***). Use of the land strongly influences the need for conservation practices and which types are most appropriate for a given tract of land.

### *Use Planning/Policies*

Land use planning involves collecting and analyzing the facts, making and carrying out decisions, and assessing results. Land capability classifications and soil surveys provide important information for planning purposes. Also to be considered are the policies of governments and social organizations and the economics resulting from using the land for an intended purpose.

### *Worn-Out Land*

Worn-out land is land that is no longer capable of producing crops at an economical level of productivity. In many cases, the condition results from much of the surface soil having been carried away during erosion by water or wind and not as a direct result of cropping. Erosion can quickly result in a worn-out soil. In contrast, soils can be kept productive for many years through proper management and protection against erosion. Through intensive management and reclamation efforts, worn-out or degraded soils can again be made productive.

## LATITUDE

The latitude of a region indirectly affects the potential for erosion by water and wind and the importance of soil water conservation. The influence of latitude stems from its effect on the climate of the region, including precipitation amount and distribution, temperature, wind speed, and evaporation potential. These effects, in turn, influence such factors as soil development, plant adaptability, soil freezing and thawing, and, thus, the potential for erosion and the possible need for water conservation.

# M

## MANAGEMENT

According to *Webster's New World Dictionary of American English,* management is "the act, art, or manner of managing, or handling, controlling, directing, etc." With regard to soil and water conservation, management, which has already been frequently mentioned, plays a major role in achieving the desired results from the implementation of a given conservation practice. For maximum effectiveness, the manager (producer, landowner, or other person) must have training and understanding of the conservation practice being used and must have an incentive to use proper management techniques. Incentives for the producer or landowner may be protecting land from erosion or conserving water for later use that could result in greater monetary gains. Incentives for others may be the reward of seeing conservation being accomplished.

### *Agroforestry*

Agroforestry is the practice of growing trees in close association with the production of cultivated field crops. Areas devoted to trees, if they are oriented across the slope of the land, serve to interrupt the continuity of the landscape, thus reducing slope length in the direction of water flow and reducing the potential for erosion by water. Potential for erosion by wind is reduced because trees provide a windbreak. Water conservation often is reduced because trees compete with crop plants for water. As a result, agroforestry is best suited for humid regions where water (precipitation) generally is adequate to supply the needs of crops and trees.

© 2006 by The Haworth Press, Inc. All rights reserved.

doi:10.1300/5678_12

### *Alternating Crop Strips*

Growing a cool-season crop such as winter wheat and a warm-season crop such as grain sorghum in adjacent narrow strips provides for water conservation when the crops are irrigated. Water conservation results from the growing crop extracting some water that remains in the soil from irrigations of the alternate crop and from use of some water derived from precipitation that falls on the area of the alternate crop. Immediate use of water derived from precipitation reduces the amount of water lost to evaporation.

### *Burning*

Burning is sometimes used to remove large amounts of residues from cropland, thereby making land preparation for the next crop easier; to eliminate old, dead grass from rangeland, thereby providing for better growth of new grasses and easier access for grazing animals; and to rid forested lands of unwanted understory plants where desirable trees are being grown. Such burning is carried out under controlled conditions. Uncontrolled burning of rangeland and sometimes of cropland results from lightning, carelessness with fires, and the deliberate actions of arsonists. Burned croplands may be highly susceptible to erosion by water and wind until they are appropriately managed. In addition, the potential for water conservation usually is decreased when crop residues are burned because such residues usually help control runoff, enhance water infiltration, and reduce soil water evaporation. The potential for greater erosion and decreased water conservation on rangeland exists until new growth of the grasses provides adequate cover of the surface. Erosion generally is not a major problem when forested lands are burned under controlled conditions because the cover provided by taller trees is not destroyed. However, it becomes a major problem when wildfires destroy all trees in a given area. Uncontrolled burning also is a serious problem on cropland and rangeland because it usually occurs during dry periods or during winter, when the chance for rapid development of new plant growth is slight.

### *Bush Fallow*

Bush fallow is the noncropped period when land is idle where shifting cultivation is practiced. Bushes, shrubs, and other plants are allowed to grow on the fallowed land. The growth of these plants may, however, be too slow to adequately protect the land against erosion when it is initially fallowed. Erosion is an especially severe problem where bush fallow is used on steeply sloping land, which is the condition where the practice is most common. Under such conditions, the amount of runoff usually is high, which results in low water conservation (*see* **CROPPING SYSTEM/SEQUENCE, *Shifting Cultivation***).

### *Cultivation of Marginally Productive Areas*

Plant growth on marginally productive areas generally is poor. As a result, not enough vegetative material may be produced to adequately cover the soil surface to provide protection against erosion by water and wind or to enhance water conservation (*see* **LAND, *Marginal Lands***).

### *Decisions*

Soil and water conservation, in general, is influenced by management decisions ranging from those of the individual producer or landowner to those of society as a whole. For most effective conservation of these resources, decisions should be based on the best information available. Those directly involved with the land must decide whether soil or water conservation is needed and whether conservation can be achieved with or without assistance. When assistance is needed, conservation specialists must decide which measures are appropriate for the given situation. Because some measures are expensive, financial assistance from a government or other source may be needed to implement an effective conservation program. For such assistance to be available, society as a whole must support the conservation efforts through programs enacted by legislative bodies or regulations developed by agencies in response to appropriate legislation. In some cases, nongovernment (private or organizational) entities support conservation efforts based on decisions made by individuals or members of different organizations.

### *Divided-Slope Farming*

Divided-slope farming is the practice of breaking up long slopes by using different tillage methods and/or planting different crops on different segments of the slope as a means of reducing the potential for erosion by water. It is similar to strip cropping (*see* **CROPPING SYSTEM/SEQUENCE, *Strip Crop Farming***) in terms of controlling erosion by water on sloping lands.

### *Early Crop Harvesting*

Crop plants use water throughout their life cycle, which may extend well past the time that their products of economic value (grain, fiber, etc.) become physiologically mature. By harvesting those products at physiological maturity and terminating the plants at that time, water use by the crop is stopped, which results in some soil water being conserved for a subsequent crop. Plowing out the plants quickly stops their water use. Fast-acting herbicides are also effective for quickly stopping water use by a crop.

### *Eppalock Reservoir Catchment Project*

The Eppalock reservoir catchment project was aimed at minimizing sedimentation that could reduce the effective life of the Eppalock Dam in Australia, for which construction began in 1960. The 2000 $km^2$ catchment area for the dam was seriously eroded. Approaches to minimizing sedimentation were constructing gully-head structures, fencing out and retiring gullied areas, and building silt traps; most of the expense for these "nonproductive" works was borne by the government. The costs of "productive" works were largely borne by landowners as they led to direct benefits to landowners through increased production. Expenses for practices such as laying out contour lines, arranging for contractors to perform certain operations, and one chisel plowing of eroded land were subsidized to some extent by the government. The project was considered a success because only one-sixth of the sediment was reaching the reservoir by the early 1970s, even though the entire catchment area had not yet been treated at that time and farm production had increased threefold in many places.

### *Farm Plan*

A farm plan is a document (management tool) prepared for the producer or landowner by personnel with an understanding of the needs and desires of the producer or landowner, potentials for erosion on the farm, and practices appropriate for controlling erosion on the land. A farm plan generally also addresses other issues.

### *Fertilization*

Good plant growth depends on the availability of adequate plant nutrients, which may require the application of fertilizer. The fertilizer may be applied as a dry, liquid, or gaseous material to the soil; as a material dissolved in irrigation water (a process known as fertigation); or, in some cases, directly to plant leaves. With proper fertilization, rapid and greater development of plant cover is possible, which reduces the potential for erosion by water and wind. Well-developed plants usually also result in sufficient plant materials after harvest that can be managed to protect land against soil and water losses until the next crop is established. Soil or plant tissue testing is highly important for determining the amounts of plant nutrients needed for the crop being grown.

### *Fire*

*See* **MANAGEMENT, *Burning*.**

### *Maintenance of Conservation Works*

Most constructed earthen conservation works are subject to natural wear and to damage by burrowing animals and animal (or even human) footpaths. As a result, maintenance and repair at suitable intervals are required to retain their effectiveness. Proper maintenance and repairs are required also for some nonearthen structures.

### *Matching New Strategies to Traditional Systems*

Tradition has a strong influence on crop production in many cases. Therefore, to more easily implement new production strategies that more effectively conserve soil or water, such new strategies should match or incorporate traditional systems to the fullest extent possible. This approach is especially important where a change is to be made by older producers or landowners whose education may be limited. Matching new strategies to traditional systems is important also under other conditions, especially when the new strategy would involve large monetary expenditures such as the purchase of new equipment. A strategy is more apt to be adopted when available equipment can be adapted for use with the new strategy.

### *Planning*

Proper planning is essential for designing and implementing appropriate practices to conserve soil and water, both where soil and water losses have been occurring and where the potential for such losses is present. Proper planning also is essential for maintaining and repairing conservation structures and practices after they have been implemented.

### *Plow-Out Practices*

A plow-out may be the conversion of rangeland to cropland to obtain more land for crop production or the plowing out of a crop that is doomed to failure because of insect or disease damage, hail, or drought, for example, thereby possibly conserving soil nutrients and water in preparation for the next crop. Appropriate, timely tillage methods should be used to minimize the potential for erosion by water and wind and to provide conditions for conserving water, when necessary.

### *Plowing to Move Soil Upslope*

On sloping land, some downslope soil movement occurs as a result of raindrop action and water movement, even when erosion by water is not a serious problem. Some downslope movement also results from normal tillage operations and may be accelerated by other tillage operations (*see* **EROSION, *Tillage Erosion***). To minimize or correct for such downslope movement, tillage, as with an implement that turns the soil in one direction, can move soil upslope. Reversible turning (moldboard) plows are satisfactory for achieving upslope soil movement.

### *Ripping*

Ripping is a mechanical method of disrupting a subsurface zone or soil horizon that restricts or prevents satisfactory water flow into a soil and/or root growth and proliferation to obtain soil water or nutrients for plant growth and production. Ripping is a costly operation and its potential for achieving the desired results (improved crop yields, erosion control, and water conservation) relative to its cost should be carefully evaluated. At some locations, a rock layer underlies the soil surface at some depth. If the rock layer is continuous and unbroken, water infiltration into the soil may be drastically reduced, depending on soil depth to the rock layer. As a result, runoff may be high, which may cause flooding, erosion by water, and limited water conservation. Fracturing the rock with an appropriate ripper can improve water infiltration and, thereby, reduce flooding, runoff, and erosion by water and improve water conservation. Deep-rock ripping is an energy-intensive, highly expensive operation and generally is limited to places where high-value crops are grown.

### *Risk*

Risk refers to the chances people are willing to take in a particular situation. With regard to soil and water conservation, the risk takers are those who more readily adopt a new conservation practice. In contrast, those with an aversion to risk generally are reluctant to adopt a new practice until it has been proven to be an improvement by others under similar conditions. Attitudes toward risk, therefore, strongly influence the rate of adoption of improved soil and water conservation practices (*see* **HUMAN FACTORS, *Age of Producer/Landowner; Attitude***).

### *Row Spacing*

Row spacing refers toward the distance between adjacent rows of a planted crop and is strongly dependent on the crop being grown. Closely spaced rows (less than ~0.30 m) are commonly used for small grain crops (e.g., wheat, barley, and oats) whereas wider spacings (0.75 to 1.0 m) are commonly used for such crops as corn, grain sorghum, and cotton. Under some conditions, narrower spacings are gaining acceptance as a result of yield increases because of improved water, nutrient, and light usage. Closer spacings also result in quicker full cover of the soil surface, thereby reducing the potential for erosion and improving conditions for water conservation. Under some conditions where precipitation is very low, row spacings of up to 2 m have been used, which allows adequate soil water for the sparsely planted crops to produce a harvestable yield.

### *Scarifying*

Scarifying is a treatment for restoring the surface of highly eroded land to a condition suitable for plant growth, thereby reducing the potential for erosion and improving conditions for water conservation. The treatment, which involves loosening or breaking the soil surface, is done with a narrow-bladed implement.

### *Seedbed Preparation*

Seedbed preparation is the process of preparing soil to promote seed germination and seedling emergence, establishment, and growth. Seedbed preparation usually involves manipulation of the soil surface by some method of tillage, which may have a positive or negative effect on soil and water conservation, depending on the method used. Tillage that results in a ridged or relatively rough (cloddy) surface or retains most crop residues on the surface provides for soil and water conservation in most cases. In contrast, tillage that leaves the surface smooth and destroys (plows under) crop residues generally is conducive to soil and water losses. Tillage that exposes moist soil to the atmosphere often increases evaporation, which is detrimental in regions where evaporation control is important for water conservation.

### *Solid Waste*

Solid wastes are the generally useless, unwanted, or discarded materials with inadequate liquid content to be free flowing. The materials may be from agricultural, commercial, industrial, institutional, municipal, or residential sources. From an agricultural perspective, any organic materials such as animal wastes, food processing refuse, waste paper, manufacturing refuse (e.g., wood chips or sawdust), and municipal and residential wastes can possibly be managed for soil and water conservation purposes. For conservation purposes, solid waste management entails the purposeful and systematic control of wastes from the point of generation to their application to land. Some solid wastes can be used to improve soil fertility, thereby improving the yield potential of crops; to increase soil organic matter content, thereby improving conditions for water infiltration and plant growth; or to provide a surface cover, thereby potentially reducing erosion and increasing water conservation. Benefits with respect to soil and water conservation from using solid wastes may be direct through their effect on surface protection, which helps control erosion and may conserve water, or indirect through improved conditions for plant establishment, growth, and production and through improved financial returns to the producer or landowner. One such material is manure, which is the fecal and urinary matter defecated by livestock and poultry. Spilled feedstuffs, bedding litter, and soil may be mixed with the defecated materials. Depending on the type of operation, collectable manure (not voided in pastures or on rangeland) may be in liquid, slurry, or solid form. Such manure can be applied to fields, pastures, or rangelands as fertilizer for improving plant growth and thereby potentially improving conditions for greater soil and water conservation.

### *Straight Plowing*

Plowing in straight lines parallel to borders often is the most convenient method of plowing a field. It is satisfactory on relatively flat land where the potential for erosion is low, but not on irregularly sloping land where strip cropping or terraces are used to reduce the potential for erosion by water. For such fields to be plowed conveniently, strips and terraces should be as parallel as possible to each other. Plowing should then be parallel to the strips or terraces, avoiding point rows as much as possible.

### *Support Practice*

For this discussion, a support practice is one that provides soil or water conservation benefits beyond those achieved with the primary practice used on the land. Some examples are using furrow diking in conjunction with ridge tillage to conserve water, using strip cropping in conjunction with contour tillage to reduce erosion on sloping land, installing drop structures in channels that convey water from terraces or other sources to more safely convey water downslope, and installing diversion terraces above a terraced field.

### *Type of Operation*

The type of production enterprise, whether field crops only, field crops and livestock, or others (orchards or nurseries), can have a strong influence on the potential for soil and water losses. Residues of some field crops (e.g., cotton) provide relatively little protection against erosion. Therefore, where such crop alone is grown, the potential for erosion may be high. In an enterprise involving crops and livestock, some land is devoted to forage production, thereby reducing the erosion potential for the enterprise as a whole. Unless adequate ground cover or other means of control are provided, erosion can be a serious problem in orchards and similar enterprises.

### *Zoning*

Zoning is the practice of partitioning a town, city, county, or other administrative region into sections to be used for different purposes such as homes, businesses, recreation, manufacturing, or agriculture. Through appropriate and careful zoning and subsequent use of appropriate conservation practices, the potential for soil and water losses can be minimized, both during the developmental stages and after completion of the different projects.

## MASS MOVEMENT

Mass movement is any downslope movement of earth surface material as a result of the direct influence of gravity. It is not a type of erosion in itself, but when flowing water is present, the loosened material may be transported from the site, which is a type of erosion. In addition, the site from which the mass of soil moved often has a high potential for erosion, especially by water.

### *Avalanche*

An avalanche is the sudden and swift downslope movement of a mass of loosened snow, soil materials, rocks, etc. from a mountain. Often, the size of the avalanche increases as it descends. The site traversed by the avalanche usually is smoothened and largely cleared of vegetation, which leaves it highly vulnerable to erosion by water until vegetation becomes reestablished.

### *Caving*

Caving is the collapse of soil, usually on a hillside and resulting from water flow into a crack or animal burrow. The collapsed soil is carried downstream by water, thus resulting in sediment deposition in streams, reservoirs, or other sites.

### *Debris Avalanches, Flows, Slides, etc.*

Debris avalanches, flows, slides, etc. are the rapid mass movements of surface materials on steep slopes in humid areas with good vegetative cover. Such movements occur during and after heavy rains, when the increased weight caused by saturation of the soil and lubrication provided by water overcome the forces that hold soil on the slope.

### *Earth Flow*

An earth flow is the downslope movement (a slip or a slide) of soil on a relatively steep hillside. The movements occur after prolonged rainy periods when the soil becomes saturated, often where downward percolation of water is impeded by subsurface conditions such as impervious rock layers or dense clays. Earth flows result in loosening of the surface soil, which may contribute to erosion by water during the ongoing or a subsequent rainstorm.

### *Landslides (Slips)*

Landslides (or slips) are the downslope movements of large soil masses on moderate to steeply sloping lands as a result of gravitational forces. They usually occur during protracted rainy periods, but the movement may or may not be related to the material being saturated with water. Soil cover may have no effect on whether a slide will occur. Where a slide occurs, bare subsoil is exposed, which may lead to increased runoff, gully formation, and erosion.

### *Mud Slides/Flows*

Mud slides or flows occur when heavy rains saturate or supersaturate soils on steep or relatively steep slopes that have been deforested or otherwise have lost their protective ground cover. The slides or flows may be rapid or slow. In the first case, the events usually occur without warning, often severely damage downslope property, and may cause loss of lives of humans and animals. Some extensive mud flows have occurred in connection with volcanic eruptions.

### *Slip*

*See* **MASS MOVEMENT,** ***Landslides*** **(*Slips*)**.

### *Slumping*

Slumping usually is a geological erosion process that can occur independently of human activities. It is characterized by a mass downslope movement of soil under high rainfall conditions on deep soils. Slumping often results in gully formation, and then continued slumping results in continued erosion as the gully advances upslope. Streambank collapse and coastal erosion also occur due to slumping.

### *Solifluction*

Solifluction is the slow downhill flow or creep of soil or other loose materials that have become saturated with water.

## MINING

Mining is the process of obtaining products such as minerals and coal from the earth, with major alterations of the land surface occurring where surface mining operations are used. Water (in some cases) and oil also are regarded as mined products, but, as for underground mining, such mining usually has little effect on land surface conditions. Where major surface alterations occur, the potential for erosion may be high and the potential for water conservation may be low. Timely land surface reconstruction (*see* **LAND,** ***Reconstruction***) can greatly reduce the potential for erosion and improve conditions for water conservation.

# MULCH/MULCHING

Mulch is a natural or artificial material placed as a layer on the soil surface to help control erosion by water or wind or to improve soil water conservation. In many cases, mulch improves microclimatic conditions, thereby improving conditions for seed germination, seedling establishment, and plant growth. Plant materials, either grown in place or transported to the location, are commonly used as mulch. Many other materials have also been used as mulch. Mulching is the process of applying mulch.

### *Bitumen*

Bitumen is a natural asphalt or a tarlike residue obtained from the distillation of petroleum or other materials. Bitumen stabilizes the soil when applied to the surface as a spray.

### *Cotton Burs/Cotton Gin Trash*

Cotton burs and cotton gin trash are trashy materials separated from lint during the ginning process for stripper-harvested cotton.

### *Excelsior*

Excelsior is used to help control erosion on slopes or in waterways while grass or other means of controlling erosion are being established. Excelsior consists of fine curly wood shavings.

### *Glass Fiber*

Glass fiber is a lightweight, foamlike mulch sprayed onto a soil surface to help control erosion, usually on steep slopes. Because it is inorganic, it is not subject to decomposition. If water runoff is reduced by applying such mulch, water conservation may be improved, which could improve plant growth on the erodible area, thereby further improving erosion control on the area.

### *Gravel*

Gravel mulch consists of relatively small stones (~5 to 10 mm in diameter) that are free of soil particles.

### *In-Row Mulching*

In-row mulching is the practice of applying mulch to the row where the crop is planted, with the space between rows remaining unmulched.

### *In Situ Mulching*

An in situ mulch is one consisting of grass or crop plant materials retained in place for the purpose of aiding in the establishment of a crop or to prevent erosion.

### *Jute Net*

Jute net is a fibrous netting made of jute that is placed on the soil surface to help establish grass on smoothed, seeded, and fertilized areas subject to erosion by water.

### *Living Mulch*

A living mulch consists of plants established, for example, in orchards to help control erosion. Periodic mowing causes litter accumulation on the soil surface, thereby improving soil and water conservation. Such plants may be annuals that produce enough seeds for annual reestablishment or perennials that do not require reestablishment.

### *Mats*

Various types of materials can be placed on the soil surface at critical points to minimize the potential for erosion, to stop erosion already occurring, and to conserve water. Materials that have been used for such purposes include mats of willow tree branches, cotton gin trash, and straw from field crops (*see* **MULCH/MULCHING, *Excelsior; Jute Net; Mulch Net***).

### *Mulch Net*

Mulch net is a loosely woven fibrous netting used to hold a mulching material such as straw or hay in place on the soil surface. The netting is anchored to the soil by wire staples. Mulch within the netting is used to prevent erosion and help establish grass on smoothed, seeded, and fertilized areas subject to erosion by water.

### *Paper*

A paper mulch consists of sheets or strips of paper. One disadvantage of such mulch is that winds may make it difficult to hold the mulch in place.

### *Paper Pellet*

Pellets made of waste paper can serve as mulch for landscaping purposes and could be used under field conditions if their cost could be reduced. Side benefits of the use of paper pellets as mulch are that it reduces the amount of paper deposited in landfills and avoids the problem with wind when sheets or strips of paper are used as mulch.

### *Pearl Millet*

Pearl millet stalks retained on the land as mulch after grain harvest have been beneficial for conserving soil and water. Millet residues generally decompose slowly, thereby providing relatively long-term benefits.

### *Plastic*

Plastic mulches, consisting of thin films of plastic placed on the soil surface, greatly reduce soil water evaporation, but provisions must be available for water from precipitation or irrigation to enter the soil and for seedlings to emerge through the mulch. By covering the ridge on which a crop is planted, infiltration is possible in furrows between the plastic-covered ridges. Slits in the plastic permit seedling emergence.

### *Residue*

A residue mulch is any mulch consisting of the stems, leaves, etc. of a crop plant. Such mulch may be retained and used at the site where the crop was grown or else harvested, transported, and placed at an offsite location. Use of residue mulches generally is an inexpensive method of controlling erosion and providing for water conservation.

### *Slot Mulching*

Slot mulching is the practice of opening a relatively narrow slot in the soil and then filling the slot with some type of crop residues (e.g., wheat straw). To be effective, the slot must remain open (not covered with soil). When installed across the slope of the land, slot mulching improves soil water conservation by reducing runoff, which also reduces the potential for erosion by water.

### *Soil/Dust*

A soil or dust mulch consists of a layer of loose soil at the surface. Such mulch generally does not directly help control erosion by water or wind. Its effectiveness for conserving water is highly dependent on precipitation distribution relative to the time when water conservation is most likely to occur. The greatest water conservation benefits occur when such mulch is established at the end of a distinct rainy season, which then results in reduced water losses due to evaporation during the ensuing dry season. Such conditions, for example, prevail in the Pacific Northwest region of the United States where winter precipitation is followed by dry summers. In contrast, a soil mulch has little potential for conserving water where the periods of greatest precipitation, potential for greatest water conservation, and greatest evaporation occur simultaneously, as in the Great Plains in the United States.

### *Stone*

A stone mulch consists of materials considerably larger than the pebbles used as a gravel mulch. Mulch of stones 50 mm or more in diameter results in greatly reduced erosion by water and, because of the reduced runoff, provides conditions conducive to soil water conservation.

### *Surface Mulch*

In contrast to a slot or vertical mulch, a surface mulch is any mulch placed as a layer on the soil surface.

### *Vertical Mulch*

A vertical mulch is similar to a slot mulch, but usually is installed to a greater depth by using a trenching machine to open the soil into which the mulching materials (usually plant residues) are placed. As for a slot mulch, the trench must remain uncovered by soil to be effective for conserving water.

# N

## NATURAL DISASTER

A natural disaster is a catastrophic event that in many cases requires large financial outlays to assist those affected and to repair or replace property damaged or destroyed by the disaster. Typical disasters are floods and storms such as hurricanes, typhoons, tornadoes, and snow and ice. Although some financial assistance for disasters may be included in budgets of responsible entities (e.g., national or other governments), assistance as a result of major disasters may exceed budgeted funds. In such cases, funds from other programs, including those dealing with soil and water conservation, may be diverted to provide relief. As a result, expenditures planned for conservation practices may become limited or unavailable, which may result in uncontrolled erosion and limited water conservation.

## NATURAL VEGETATION/ NATIVE CONDITION

The terms "natural vegetation" and "native condition" imply something that has always existed on a given tract of land. Although such vegetation or conditions have existed for a long time, changes may have occurred because of climatic changes and catastrophic events (e.g., floods, drought, or fires) that resulted in an invasion by different plant species or changes in land surface conditions (e.g., stream channel changes, gully formation, or dune formation). Natural vegetation

© 2006 by The Haworth Press, Inc. All rights reserved.
doi:10.1300/5678_13

and native conditions for this discussion refer to those that existed when humans first settled in a given region and have not been altered by humans since that time. The vegetation and other conditions that existed in the Great Plains of the United States, for example, when the region was settled by humans were such that erosion by water and wind apparently was not a serious problem. Human alteration of the soil surface by plowing out the prevailing grasses to grow crops, however, contributed to the severe erosion by wind that occurred in the region in the 1930s. Areas still under natural vegetation or native conditions in many cases remain highly resistant to erosion. Natural vegetation and native conditions in some other regions, however, were not as effective for controlling erosion, as indicated, for example, by the severely eroded landscapes of the Badlands in South Dakota, the Painted Desert in Arizona, and Monument Valley in Arizona and Utah in the United States.

# P

## PERCOLATION

Percolation is the usually rapid penetration of water in a soil to depths beyond the rooting depth of most plant species. Percolated water, therefore, becomes unavailable for plant use and reduces soil water storage (conservation) for later use by plants at the point of percolation. Some percolated water may become available at a downslope seepage site, in a spring, or in a stream that could be used for irrigation.

## PERENNIAL GRASSES AND OTHER VEGETATION

Perennial grasses and other types of perennial vegetation are those whose life cycle is greater than 2 years. Such plants regrow from their root systems year after year, thus usually being highly effective for helping control erosion on areas highly susceptible to erosion and for conserving water. Because annual reestablishment is not required, such vegetation provides constant control throughout the year, which may not be possible where land must be prepared and the vegetation reestablished each year.

© 2006 by The Haworth Press, Inc. All rights reserved.

doi:10.1300/5678_14

# PESTS

Crop, pasture, and range plants and the land on which they grow are subject to damage or destruction by a variety of pests, including insects, plant diseases, and marauding animals. Because of such damage or destruction, the potential for erosion may be increased and the potential for water conservation may be decreased. Some pests may also damage or destroy structures installed for soil or water conservation purposes.

## *Animals*

Numerous animals live in the soil and others live in fields or areas adjacent to fields (e.g., pastures and rangelands). Animals living in soil may feed on plants (roots or aboveground parts), which may reduce the protection of the land needed for controlling erosion and conserving water. In addition, burrowing by such animals may damage terraces or waterways, thereby resulting in accelerated erosion at localized points in a field. Other animals living in fields may also feed on crops, and some (e.g., deer, wild hogs, wolves, or raccoons) may enter fields when satisfactory forages or other feed materials are not available outside the field. Such animals sometimes cause extensive damage to crops, thereby increasing the potential for erosion, reducing the potential for water conservation, reducing yields, and reducing income for the producer or landowner. In addition to the direct effect on water and soil conservation, such damage may indirectly lead to failure to implement necessary soil or water conservation practices owing to reduced income for the producer or landowner.

## *Insects*

Insects affect soil and water conservation through their effect on plant establishment, growth, and yield, and on potential carryover of crop residues that could be managed for controlling erosion and enhancing water conservation. Insects such as grasshoppers, army worms, and termites sometimes devour most leafy vegetative materials on the landscape in a given area, thus leaving soils highly vulnerable to erosion by water and wind and reducing the potential for water conservation. Reduced yields resulting from insect damage may, in turn, lower income for the producer or landowner, which may have an impact on whether conservation practices are used on the land. Insect control, therefore, is an important crop management activity. Insect control is possible by using appropriate insecticides, insect resistant cultivars, and integrated pest management practices. The latter rely on controlling the climate, food sources, and entry points to prevent and manage pest infestations. Chemicals are used only in crisis situations that threaten rapid losses or when pests are not controlled by the more conservative methods.

### *Nematodes, Mites, etc.*

Other organisms that damage or destroy plants include nematodes and mites, and appropriate control measures should be used to maintain the vigor of plants, thereby achieving satisfactory plant growth, cover, and productivity, which are important for protecting soil against erosion and for conserving water.

### *Plant Diseases*

Most plants are subject to some disease. A plant disease may completely destroy a crop, thereby leaving the soil surface bare and susceptible to erosion by water and wind. If the condition results in increased runoff, the potential for water conservation is decreased also. If the crop is damaged, only limited amounts of surface residues may remain, which also can increase the potential for erosion by water and wind and decrease the potential for water conservation. Potentially greater erosion and lower water conservation may also occur if the damaged crop is destroyed (e.g., plowed under) to eliminate the diseased plant materials (*see* MANAGEMENT, ***Plow-Out Practices***).

### *Weeds*

Weeds generally are detrimental to crop production and water conservation because they compete with crop plants for water, nutrients, space, and light. Weeds, however, can provide protection against erosion during the interval between crops when other means for controlling erosion are not available. This could be the case when crop residues are not available (harvested or used for other purposes) or other suitable erosion control methods cannot be established (delayed because of unfavorable weather, wildfires, etc.). In such cases, allowing nonnoxious weeds to grow until they provide adequate cover on the surface can reduce the potential for erosion. When adequate growth occurs, the weeds can be terminated using tillage or herbicides, with their residues remaining on the surface as a mulch for continued protection against erosion and for water conservation purposes. Although weeds can provide for soil and water conservation under some conditions, their control at some time generally is essential for successful crop production.

*Herbicides*

The possibility of controlling weeds with herbicides has resulted in reducing or eliminating tillage for crop production in many cases. When weeds are controlled with herbicides, the soil is not disturbed by cultivation, which may reduce the potential for erosion by water and wind and conserve water by reducing evaporation from moist soil exposed to the atmosphere by tillage. Also, most crop residues can be retained on the soil surface, which is effective for reducing the potential for erosion by water and wind and enhancing the potential for water conservation. Further water conservation is achieved by controlling weeds because they compete with crops for water. Weed control methods with herbicides include:

1. Broadcast applications when weed control on the entire field area is needed, with such applications being made to growing weeds during the interval between crops, to soil with subsequent incorporation by tillage for control during a crop's growing season, or to an established crop by using a herbicide that controls the weeds and does not harm the crop.
2. Directed spray, which involves weed control in established crops by targeting the weeds and shielding the crops from the material being applied using appropriate equipment.

3. Spot control, which involves the use of spray equipment with sensors to detect weeds and activate the equipment to apply the herbicide only to detected weeds. Spot control results in using lower amounts of herbicides and, therefore, is beneficial for economic and environmental reasons. Hand-operated equipment can also be used for spot weed control.
4. Ultra-low volume (ULV) spray, which consists of an extremely fine mist of the carrier material containing the herbicide that penetrates all spaces around the plant and greatly reduces the amount of herbicide applied. The ULV spray method is also used to apply insecticides.

Weed resistance to herbicides has occurred under some conditions, which may necessitate a change to other currently available herbicides, development of new herbicides, or a return to greater use of tillage for controlling the weeds. The latter possibly would increase the potential for erosion by water and wind and reduce the potential for water conservation.

### *Flame Cultivation*

Flame cultivation is the practice of killing small weeds, usually in established crops, with controlled and directed flames. The effect on erosion by water and wind is indirect. By eliminating mechanical tillage to control weeds, soil is disturbed less, which may render it less subject to movement by flowing water. Erosion by wind should be little affected by flame cultivation because, by the time plants are large enough to tolerate flame cultivation, their growth should be sufficient that erosion by wind is no longer a problem.

### *Herbicide-Tolerant Crops*

Some herbicides are available to control certain types of weeds without damaging a given crop. The same herbicides also eliminate unwanted plants of another crop (volunteer plants) along with the weeds. Control of all weeds without damaging the crop would be the ideal situation. Such weed control has become possible by genetically altering crop plants. When such plants are grown, the herbicide can be applied to the growing crop plants whenever weed control becomes necessary.

*Tillage*

Most tillage methods have some effect on controlling weeds. In general, tillage that mixes or inverts the surface soil layer (e.g., disk implements and moldboard plows) is more effective for controlling weeds than tillage that undercuts the soil surface (e.g., stubble mulch tillage).

## PHREATOPHYTE

Phreatophytes are water-loving plants that derive their water from subsurface sources and usually are of no economic value. Such plants often grow beside streams, canals, and other waterways, from which they derive their water. Some deep-rooting phreatophytes derive their water from water tables. Water use by such plants reduces the amount available for other purposes (agricultural, residential, industrial, and recreational).

## PLANT

A plant is any aboveground organism able to carry on photosynthesis in its cells, which contain chloroplasts and have cellulose in the cell walls. For this book, all field crops, all pasture and rangeland plants (including grasses, forbs, shrubs, bushes, and trees), and forest trees are of particular interest. A wide range of plant types occur in nature or are used in productive systems. Soil and water conservation is strongly dependent on the species present at a given location. Plant characteristics and management practices that have a direct or indirect effect on soil and water conservation include the following.

### *Adaptability*

For optimum productivity and benefits with respect to soil and water conservation, plants that are well adapted to the given location should be selected when a choice is possible, as for field crops, pasture forages, and trees.

### *Growth Habit*

Each plant species has a characteristic shape or form that results, for example, from its branching mode, twig characteristics, foliage type, top form, flowering and fruiting form, and bark type. Plant species resulting in a high stem number or cover at the soil surface are highly effective for controlling erosion, both by water and wind, and for improving water conservation because of reduced runoff. Control of erosion by wind is possible using any plants that reduce wind speeds below threshold values at the soil surface.

### *Growth Stages*

The potential for soil and water losses generally is greatest at the time of plant establishment and before adequate growth occurs to provide protection against raindrop impact, water flow, and wind forces. In some areas, the potential for erosion is high after crop harvest, as for cotton on sandy soils, because the stalks provide little resistance to flowing water or wind. Soil and water losses may also be high in the period between successive crops (before crop establishment).

### *Rooting Depth and Type*

Plants that root deeply, and especially those with a fibrous root system, provide maximum protection against the potential for erosion by water.

### *Spacing*

Closely spaced crop plants (e.g., small grain crops) provide greater resistance to water flow across the soil surface than crops generally planted in more widely spaced rows or at lower populations within rows (e.g., corn and cotton). Closely spaced plants, therefore, generally provide greater protection against erosion by water and improve the potential for water conservation. Plant spacing generally has little effect on erosion by wind, except possibly at an early growth stage, after harvest, or when plant spacings are extremely wide.

## PLANTER/SEEDER TYPE

The effect of planter or seeder type on soil and water conservation occurs through its effect on crop establishment and surface conditions resulting from the planting operation. When conditions such as soil water content and temperature become favorable, using a planter best suited for the crop provides the best opportunity to rapidly establish the crop, which then more quickly results in adequate plant growth to provide soil and water conservation benefits. The type of planter used affects soil surface conditions at the time of and immediately after planting. Some planters result in a relatively smooth surface, whereas others result in ridges and furrows or other forms of roughness (surface depressions), which may provide protection against erosion by water and wind and improve water conservation.

## PLANTING/SEEDING

A number of factors related to crop planting or seeding directly or indirectly affect the potential for soil and water conservation on land where the operation is performed.

### *Coulters*

Planters or seeders equipped with coulters ahead of the seed spout allow planting where significant amounts of residues remain on the soil surface. Crop establishment where residues remain on the surface improves conditions for soil and water conservation. Use of coulters is essential for crop establishment where conservation tillage, especially no-tillage, is practiced.

### *Critical Areas*

Some field areas are more prone to erosion and have a greater need to conserve water than others. As a result, a special planting or seeding method, crop, etc. may be needed to minimize the potential for soil and water losses.

### *Deep (In-Furrow) Seed Placement*

Soil water conditions at normal seed placement depths sometimes are not satisfactory for seed germination and seedling establishment at the desired time, but better conditions may exist at a greater soil depth. Using a planter or seeder capable of opening soil to the greater depth, placing the seed, and covering the seed to a normal depth allows timely crop establishment and greater potential for soil and water conservation than if crop establishment is delayed until conditions become favorable at the normal depth (*see* **PLANTING/SEEDING, *Methods,*** *Lister Planter*).

### *Depth of Seed Cover*

Each seed type has a depth to which it should be covered with soil to achieve optimum germination and seedling establishment. With proper depth of seed placement, plant establishment occurs more readily, thus providing for plant cover in a timely manner and, thereby, providing conditions for controlling erosion and conserving water.

### *Direct Drilling/Seeding*

*See* TILLAGE, ***No-Tillage.***

### *Methods*

Crop seeding can be achieved by several different methods.

#### *Drills*

Drills are commonly used for seeding small grain crops such as wheat, barley, or oats, and for grasses. Different drills have disk, hoe, or shoe openers to open the soil for seed placement, with the seed being covered by a drag or other device and firmed with an appropriate press wheel. Under many conditions, ridges formed by openers provide adequate surface roughness to control erosion until the crop becomes established.

#### *Lister Planter*

A lister planter is a seeding unit that opens a relatively deep furrow in which the seed is then planted at an appropriate depth. This method is sometimes used for dryland crops when soil water conditions near the surface are not suitable for timely crop establishment. When used on a contour, lister planting provides for control of erosion by water and water conservation benefits (*see* **PLANTING/SEEDING, *Deep [In-furrow] Seed Placement***).

#### *No-Tillage*

*See* TILLAGE, ***No-Tillage.***

#### *Plow-Plant Planting*

Plow-plant is the practice of plowing a field and planting a crop in a single trip across the field with one tractor suppling the power for both operations. Because plowing is delayed until planting, the potential for erosion before planting is reduced, especially if crop residues, a cover crop, or other vegetation is on the land.

### *Ridge Planting*

Ridge planting is the practice of planting crops on ridges formed during previous tillage or seedbed preparation operations. Commonly, only one seed row is planted on each ridge, but two seed rows are planted on each ridge for some crops (e.g., grain sorghum in some areas) and several seed rows are planted on each ridge for drill-planted grain crops (e.g., winter wheat). Ridge planting when done on the contour of the land can provide for control of erosion by water and water conservation. When properly oriented perpendicular to wind direction, ridge planting can provide for control of erosion by wind.

### *Slit Planting (Slot Planting)*

Slit planting is a method of planting whereby a narrow slit is opened, usually with a coulter attached to the planter, to place seed in soil at the desired depth. No prior seedbed preparation is involved and herbicides are applied shortly before, during, or soon after planting when such planting is used in reduced- or no-tillage systems (*see* **TILLAGE, *No-Tillage***).

### *Strip-Till Planting*

Strip-till planting is the practice of planting a crop in a narrow (30 to 50 cm wide) seeding zone prepared by rotary tillage (or other implement) through a living mulch (cover crop) or residues from the previous crop. During planting or soon afterward, the remainder of the living mulch is cut loose (undercut), killed, or retarded. Retaining some of the mulch or crop residues on the surface provides erosion control and water conservation benefits.

### *Wheel-Track Planting*

Wheel-track planting is the practice of planting seeds in tracks formed by, for example, tractor wheels immediately ahead of the planting unit. Wheel-track planting usually leaves much of the land surface in a relatively rough condition, thereby reducing the potential for erosion by water and wind and improving the potential for water conservation.

### *Point Rows (and Planting Across Row Ends)*

Point rows usually occur in fields with irregular boundaries or where the rows follow the contour of the land. To maximize the planted area in such fields, crops sometimes are planted across the row ends. Planting across row ends sometimes also occurs when large planting equipment is used, which results in a wide turning area. Planting across row ends on sloping land usually results in those rows being oriented with the slope and, therefore, may result in major soil or water losses at such places in a field.

### *Rainfall Pattern*

Each cultivar of a crop species provides best results when planted within a relatively narrow time frame because of day length, growing degree days (total hours of air temperatures above a certain minimum), and possibility of cold (freezing) temperatures before harvest. These factors must be considered when selecting a cultivar. Also to be considered is the potential for timely precipitation when irrigation is not an alternative for supplying water to the crop. When conditions for seeding are not suitable at the desired time, good yields often are still possible by switching to an alternative cultivar when conditions become favorable (e.g., switching from a long-season to a medium- or short-season cultivar). A short-season cultivar may result in less plant growth than a long-season cultivar, but any plant material produced usually provides for greater soil and water conservation than where no crop is grown.

### *Rate/Plant Population*

All crops have an optimum plant population for achieving optimum production under conditions prevailing at a given location. To achieve desired populations, use of proper seeding rates is essential.

To determine seeding rates, the seed germination percentage must be considered, and using treated seed to minimize the potential for damage by insects usually is beneficial. Optimum plant populations usually also provide conditions favorable for soil and water conservation.

### *Seed Cradle*

Seed cradles are holes made in soil in which crop seeds are planted. The holes are about 10 to 15 cm deep and 10 to 15 cm in diameter and provide for water accumulation. They also provide a protective environment by partially shading the plants and by reducing evaporative water losses and sandblasting by wind.

### *Skip-Row Cropping*

Skip-row cropping and planting is the practice of not planting the crop on all rows in a field. It is sometimes used for dryland (nonirrigated) crops in semiarid regions. Typical arrangements are two rows planted, one row not planted and two rows planted, two rows not planted. Such cropping patterns provide some additional water to the crop on the planted rows.

### *Slot Planting*

*See* **PLANTING/SEEDING,** ***Methods,*** *Slit Planting (Slot Planting)*; **TILLAGE,** ***No-Tillage.***

### *Staggered Planting Dates*

Using staggered (or several) planting dates for a given crop improves the potential to achieve some production from a crop that is highly dependent on, for example, precipitation at a critical growth stage. This practice is mainly appropriate for dryland crops in a semiarid region where precipitation is limited and highly erratic. Achieving some production provides some income for the producer or landowner, thus potentially providing financial resources for implementing or maintaining soil and water conservation practices.

## *Strip Tillage*

Strip tillage and planting is the practice of planting crops in strips of varying width, with the intervening area often having residues from the previous crop in place and being left in an undisturbed condition. Strip cropping is widely used for controlling erosion by wind, for example, in portions of the Great Plains of the United States and the Prairie Provinces of Canada.

## *Time*

Timely planting is essential for optimum crop productivity. Where seedbed preparation before planting is practiced, it should, if possible, be done in a timely manner so that planting can be carried out at the optimum time. The plow-plant and no-tillage planting methods provide some opportunity for more timely planting than methods that require major seedbed preparation before planting. This is especially the case when precipitation and, therefore, wet soil conditions may delay seedbed preparation. Timely planting is important for rapidly obtaining plant cover on the land, which is important for controlling erosion by water and wind.

## *Transplanting*

Transplanting is the process of establishing plants (e.g., seedlings, trees, shrubs, and rhizomes [some grasses]) at a different site from the one where the materials were originally grown (in a nursery, hothouse, or field). Transplanting is widely used to reestablish trees in forests, establish tree or shrub windbreaks at home sites or other places, and establish some grasses in pastures. Transplanting provides for relatively quickly establishing plants that provide for protection against erosion and may also provide for water conservation.

## PRODUCTION LEVEL

The production level of a crop is largely affected by plant growth. That is, poor growth results in low production and good growth usually results in relatively good production. The production level, therefore, also affects the potential for soil and water conservation. With poor growth, land may not be adequately protected while the crop is in place and insufficient residues may be produced to protect land after the crop is harvested. With good growth and adequate residue production, residues can be managed for soil and water conservation during the noncrop period. Another effect of production level on soil and water conservation is through the financial returns received by the producer or landowner. With good returns, the potential for investing in conservation practices is greater than when returns are limited, as with a poor crop production level.

# R

## RANGELANDS/GRASSLANDS/PASTURES

Rangelands, grasslands, and pastures are tracts of land managed for forage production, with the forage usually harvested by grazing by domesticated animals or wildlife. In some cases, forages may be harvested as hay for later feeding of animals. The forage may be native or introduced grasses, legumes, forbs, shrubs, and so on. Proper management of these lands is essential to achieve optimum productivity, to minimize the potential for erosion, and to conserve water for optimum plant growth. Soil or water conservation on some lands may be good or poor depending on how they are managed.

### *Carrying Capacity*

The carrying capacity of a range refers to the maximum number of animals that the range can support without damaging the vegetation. Fluctuating forage production because of precipitation differences and other factors causes the carrying capacity to differ from year to year. The carrying capacity should include wild animals on ranges where they may compete with domestic animals for forage. Frequently exceeding the carrying capacity generally leads to poor growth of the vegetation and degradation of the range, which increases the potential for soil and water losses.

### *Controlled Burning*

Controlled range burning is practiced in some regions, with the goal of removing dead, unpalatable vegetation from the previous

© 2006 by The Haworth Press, Inc. All rights reserved.
doi:10.1300/5678_15

year's growth and making new growth more easily accessible to grazing animals. Burning also helps control some unwanted plants (e.g., weeds and brush). Burning at the proper time generally does not lead to serious soil and water losses (*see* **MANAGEMENT, *Burning***).

### *Grazing*

For an established pasture or rangeland, appropriate grazing management is a prerequisite for achieving optimum productivity and minimizing the potential for soil and water losses. Through grazing management, which includes maintaining a proper balance between animal populations and forage availability, adequate plant cover is kept on the land for erosion control and water conservation. Techniques involved in grazing management include deferred and rotation grazing. For deferred grazing, animals are not allowed to graze until the forage plants become well established or reach a desirable growth stage. Deferred grazing can also be used to allow plants to recover after a period of relatively intense grazing on a given area. Rotation grazing involves allowing grazing on a given area for a relatively short time, then allowing grazing on each successive area also for a short time before again allowing grazing on the first area.

### *Pitting/Ripping/Cutout Disking/Furrows on Contour*

Range pitting, ripping, cutout disking, and furrowing on the contour sometimes are used to trap runoff water, thereby providing more time for water infiltration and increasing soil water storage. For cutout disking, a part of each disk is removed, resulting in a series of depressions being cut in the soil surface instead of continuous cutting (and mixing) as the implement is pulled across the land surface (*see* **MANAGEMENT, *Ripping***).

### *Prairie Grassland*

Prairie grassland is the natural vegetation that developed, for example, in the Great Plains of the United States, where the climate involves generally low precipitation and severe droughts periodically occur. As for other grasslands, good management is needed to maintain productivity and to achieve erosion control and water conservation when prairie grasslands are grazed by animals.

### *Renovation*

Renovation of grasslands that have become degraded (e.g., as a result of overgrazing or drought) can restore them to a condition that reduces the potential for subsequent erosion by water and wind. Some surface shaping, deep loosening, or other practices may be required for successful renovation of severely degraded grasslands that have active gullies or exposed subsoil horizons. Successful renovation also requires the establishment and subsequent proper management of a grass well adapted to the prevailing soil and climatic conditions of the site (*see* **RANGELANDS/GRASSLANDS/PASTURES, *Revegetation; Seeding***).

### *Reseeding*

Under some conditions, pasture productivity declines and reseeding may be required to reestablish the original species or to introduce a new species. Reseeding may be required because of insect damage, plant disease, or overgrazing. When reseeding is chosen, proper land preparation should be carried out before seeding and the new forage plants should be allowed to become well established before grazing is resumed to minimize the potential for erosion and to achieve water conservation.

### *Rest Period*

A rest period is a time during which grazing animals are restricted from a range or pasture to allow existing grass species to develop or reseed for the purpose of achieving improved forage production under subsequent grazing conditions. Through increased growth and reseeding of the grass, the improved range conditions provide for better erosion control and water conservation. A rest period is also a natural period of dormancy for a plant species during which plants do not increase in size even though conditions for growth are favorable.

Depending on plant size, conditions for soil and water conservation may or may not be favorable during a rest period.

### *Revegetation*

Revegetation is the practice of reestablishing a forage species that has lost productivity for some reason or of replacing the existing species with a different species that may be more suitable for forage production at the given location. Revegetation can be accomplished through seeding (*see* **RANGELANDS/GRASSLANDS/PASTURES, *Seeding***) or by planting stolons, as for establishing Coastal Bermuda grass. Stolons are stems that grow horizontally along the soil surface and produce roots and top growth at the nodes. With proper management, revegetation usually provides for improved conditions to conserve soil and water.

### *Rotation Grazing*

*See* **RANGELANDS/GRASSLANDS/PASTURES, *Grazing.***

### *Salt/Mineral/Feeder Location*

Animals have a tendency to graze at the point where desirable forage first becomes available. If salt, other minerals, or even some supplemental feeds are provided, the animals often visit those places and resume grazing nearby, potentially leading to overgrazing at those places. Improved grazing distribution can be achieved by locating salts, minerals, and feeders at several locations in the pasture or on the rangeland, or by periodically moving them to different sites.

### *Seeding*

Seeding of a forage species on a range usually requires that a special drill be used. Some species can be seeded directly into the existing vegetation, whereas others require that a suitable seedbed be prepared for optimum establishment. Timing of range seeding is critical for achieving satisfactory grass establishment, for minimizing competition with the existing vegetation, and for achieving satisfactory erosion control and water conservation.

### *Stocking Rate*

*See* **RANGELANDS/GRASSLANDS/PASTURES, *Grazing*.**

### *Water Supply*

As for salt, minerals, and supplemental feeds, animals regularly visit places where water is available, which can lead to overgrazing near the water sources when the animals resume grazing nearby. Providing water at several points in the pasture or on the range provides for better grazing distribution, which, in turn, reduces the potential for erosion near the water sources. Water can be provided at several places from ponds (dug in the soil), natural springs, streams, or wells or through pipes with water supplied from another site and discharged into a trough.

## RESIDUES

From an agricultural perspective, residues are the materials remaining on land after harvesting the crop for its grain, fiber, forage, etc. or any materials from an industrial operation that may be suitable and available for distribution on land.

### *Crop Residues*

Crop residues are those parts that remain in the field after harvesting the crop for its grain, fiber, forage, etc. Factors or characteristics of crop residues that affect soil and/or water conservation include the following.

*Amount*

The amount of residues available is probably the most important factor related to soil and water conservation. With complete cover of the surface, erosion by water and wind is usually negligible or avoided, and the potential for water conservation is good. As residue amounts decrease, the potential increases for erosion and decreases for water conservation. A 30 percent surface cover is about the minimum for achieving satisfactory erosion control under most conditions. This amount of cover usually also provides some water conservation benefits.

*Anchorage*

For maximum effectiveness in erosion control and water conservation, crop residues should be securely anchored in the soil to avoid being carried away by water or wind and to maintain a relatively uniform distribution on the land surface.

*Incorporation (Mixing with Soil)*

Although surface residues generally are most desirable for conservation purposes, residue incorporation has benefits for conservation purposes on some soils. The addition of organic materials to the soil helps to reduce soil compaction, provides temporary depressions or voids for water storage, and helps improve soil structure, for example. Crop residues are incorporated with soil where clean tillage is practiced. Residue incorporation on some soils and under some conditions may result in the soil being vulnerable to erosion by water or wind and or having a low potential for water conservation (*see* **TILLAGE, *Clean Tillage/Clean Culture***).

*Irrigation Furrows*

Crop residues in irrigation furrows help control erosion when irrigation water is applied in some cases, but impede water flow in other cases to the point that water distribution in the field becomes highly irregular. Under such conditions, tillage can be used to remove residues from the furrows after crop establishment and before applying the first irrigation. Retaining residues until this time provides for good protection of the surface for soil and water conservation purposes at the time of crop establishment.

*Management*

Management in this case refers to managing those crop residues that remain after harvest. The residues may be retained in place for soil or water conservation purposes; removed, transported, and applied for conservation purposes at another site; plowed under or burned to prepare the land for the next crop (*see* **RESIDUES,** ***Crop Residues,*** *Incorporation [Mixing with Soil]*); or removed for other purposes (e.g., feed or bedding for animals, fuel, or building materials; *see* **RESIDUES,** ***Crop Residues,*** *Use as Fuel/Other Purposes*).

*Orientation*

Provided adequate amounts are present, residue orientation (e.g., standing as opposed to flattened on the surface) usually has relatively little or no effect on erosion by water or wind. With relatively sparse residues, however, the potential for erosion by water may be greater with standing residues because more raindrops may impact the soil surface, thereby increasing aggregate dispersion, surface sealing, and runoff, which also reduce water conservation. Erosion by wind should not be affected, unless the residues are extremely sparse. Residue orientation affects water conservation through its effect on runoff, infiltration, and evaporation, with each being strongly affected by the amount present. In addition, evaporation is a highly complex process affected by soil temperatures, residue reflectance, wind speed at the surface, and so on. In relation to evaporation, water conservation may, therefore, be greater with flattened residues in some cases and greater with standing residues in other cases. Water conservation resulting from snow trapping is greater with standing than with flattened residues on the soil surface.

*Removal*

*See* **RESIDUES, *Crop Residues,*** *Use as Fuel/Other Purposes.*

*Surface Residues*

*See* **RESIDUES, *Crop Residues,*** *Amount.*

*Stubble Orientation*

*See* **RESIDUES, *Crop Residues,*** *Orientation.*

*Type*

The type of residue produced by a crop strongly affects the potential for conserving soil and water. For example, crops such as small grains that result in numerous stems close to each other are much more effective for controlling erosion by flowing water than those such as cotton and sunflower that result in more widely spaced stems. In addition, crops with a hollow stem (e.g., winter wheat) provide more surface cover on a same-weight basis than those with a pithy stem (e.g., corn or grain sorghum) or a solid (woody) stem (e.g., cotton). Greater surface cover improves conditions for soil and water conservation. Again, on a same-weight basis, the lighter (less dense) residues result in a thicker residue layer on the surface, thereby being more effective for reducing evaporation than a same-weight layer of more dense residues.

*Use as Fuel/Other Purposes*

When adequate amounts are available, proper management of crop residues is effective for controlling erosion by water and wind and for improving water conservation. Under some conditions, however, crop residues are used as fuel for heating and cooking, animal feed and bedding, building materials, etc., which then reduces or negates their potential use for soil and water conservation purposes. Where residues are required for other purposes and soil and water conservation is also needed, efforts should be made to remove only a part of the residues or the part most desirable for the given use, leaving the remainder on the land for conservation purposes. Other means of reducing dependence on residues for feed, fuel, and so on are to substitute high-value crops for residues, use alley cropping to obtain alternative feed and fuel materials, utilize wasteland areas for growing forage and fuel materials, establish an improved balance between animal populations and feed supplies, and use alternative fuel sources.

### *Industrial Residues*

Industrial residues are waste materials derived and generally discarded from industrial operations or manufacturing processes. Depending on their nature, some industrial residues can be used directly or with some processing to achieve soil or water conservation. Such materials include by-products of food processing (e.g., hulls and peels), wood or paper from a manufacturing process, tree bark, sawdust, and wood chips.

## RUNOFF

Runoff is that portion of precipitation or irrigation water that does not infiltrate soil and flows across the soil surface into channels or streams. Runoff is one transport agent for waterborne sediments carried from land when erosion occurs (splash erosion also transports sediments). Water conservation also is reduced when runoff occurs. The potential effect of runoff on erosion by wind is indirect. Runoff may smoothen the soil surface, thereby making it more susceptible to erosion by wind. In addition, when water conservation is reduced, plant establishment and growth may be detrimentally affected, which could result in land being more susceptible to erosion by wind. Although runoff may result in erosion by water and limited water storage in soil, it is the source of water stored in reservoirs and, therefore, provides some water for nonagricultural users.

# S

## SEASON

In most regions, the greatest potentials for erosion by water and wind occur in a particular season of the year. Likewise, the potential for achieving water conservation is also highly seasonal. Therefore, for optimum soil and water conservation, it is highly important that effective practices be in place at the most critical season of the year to achieve the greatest conservation benefits.

## SEDIMENTATION

Sedimentation is the process of sediment deposition. Sediment deposition in a channel or stream reduces the capacity to carry water and, therefore, could result in flooding and/or erosion by water of adjacent lands when, for example, heavy rains occur. Sediment deposition decreases the water storage capacity of reservoirs, thereby potentially reducing the amount of water available for useful purposes (agricultural, residential, industrial, recreational).

## SEDIMENT-RETENTION BASIN

A sediment-retention basin is a pond or basin in a reservoir or channel in which sediment-laden water is stored long enough for the

© 2006 by The Haworth Press, Inc. All rights reserved.
doi:10.1300/5678_16

sediments to settle from the water. Retaining such sediments avoids their deposition in a channel, which would decrease its water flow capacity, or in a reservoir, which would decrease its water storage capacity.

# SOIL

Soil is the layer of unconsolidated mineral and organic materials at the immediate surface of the earth that serves as the natural medium for the growth of land plants. Many characteristics and conditions of a soil influence its susceptibility to erosion and water losses and, therefore, how it must be managed to minimize or avoid such losses. The influence may be direct or indirect through its influence on plant establishment, growth, and yield, which may affect, for example, development of plant cover, crop residues produced, and financial returns to the producer or landowner.

## *Additive/Amendment*

An additive or amendment is any substance or treatment imposed on a soil to make it more suitable for plant growth and/or less susceptible to soil and water losses.

### *Emulsion*

An emulsion is any stable material suspended in a suitable liquid. Emulsions of oil or other petroleum products applied to sandy soils have reduced their susceptibility to erosion by wind. Such applications generally are suitable for controlling erosion on relatively small areas such as those around work locations, factories, and power plants, and not for large-scale field areas. For soil and water conservation, an emulsion may be a water-based mixture containing oil, latex, or bitumen. The emulsion is sprayed on a mulch of fine cellulose fibers placed on the soil surface to accelerate the establishment of vegetation on sloping land. Grass seed and fertilizer usually are applied with the mulch and/or emulsion. The emulsion forms a thin layer on the mulch surface, holding the mulch particles in place. The emulsion does not restrict seed germination and may even improve it.

*Gum Arabic*

Gum arabic is a gum derived from some species of Acacia trees. It is used to stabilize some emulsions, such as those used to treat sandy soils to help control erosion by wind.

*Lime/pH*

All plants grow best when the soil pH is at the optimum level for the particular plant species. When pH is sufficiently below the optimum, a particular species will grow poorly or not at all. As a result, plant cover on the soil surface will be low or nonexistent, which could subject the soil to erosion by water and wind and decrease the potential for water conservation. In these circumstances, applying lime increases the pH of a soil, thus improving conditions for growth and, thereby, increasing the potential for soil and water conservation.

*Phosphogypsum*

Phosphogypsum is a by-product derived when sulfuric acid reacts with phosphate rock to produce phosphoric acid, which is necessary for the production of phosphate fertilizers. Phosphogypsum applied to soils with different textures (clay loam, silty clay loam, loam, or fine sandy loam) results in greater water infiltration than when soils are not treated. Infiltration is greater because the phosphogypsum stabilizes the aggregates at the soil surface. Greater infiltration increases the potential for greater water conservation and, therefore, lower erosion by water. Currently, use of phosphogypsum is restricted in the United States because of environmental concerns.

*Polyacrylamide*

Polyacrylamide (PAM) is an anionic polymer. Injecting PAM or starch copolymer solutions into water used for furrow irrigating crops reduces runoff, thereby increasing water infiltration and conservation and reducing surface sealing and soil sediment losses.

### *Aeration*

All plants grow best when the aeration status of a soil is at the optimum level for the particular plant species. When the aeration is sufficiently below the optimum, a particular species will grow poorly or not at all. As a result, plant cover on the soil surface will be low or nonexistent, which could subject the soil to erosion by water and possibly by wind and decrease the potential for water conservation. Poor aeration results from poor soil drainage or possibly from soil compaction. Providing for adequate drainage of excess water or loosening compacted soils (e.g., by tillage) improves conditions for growth of the particular plant species and, thereby, reduces the potential for erosion and increases the potential for water conservation on compacted soils.

### *Aggregate Stability*

The stability of surface soil aggregates strongly affects the potential for soil and water conservation. Dispersion (breakdown or disintegration) of low-stability aggregates occurs during rainstorms as a result of raindrops impacting the soil surface. Fine soil particles resulting from the dispersed aggregates move with water and clog soil pores, thus causing the formation of surface seals that increase runoff. This reduces infiltration and water conservation and increases the potential for erosion.

### *Aggregation*

Soil aggregation is the cementing or binding together of several to many soil particles into a secondary unit or aggregate. Aggregates that are stable in water are highly important for maintaining a soil structure that is conducive for water infiltration and plant growth. Unstable aggregates disintegrate when wetted, thereby resulting in surface sealing, increased runoff, and potentially greater soil and water losses.

### *Aluminum Toxicity*

Aluminum toxicity is a problem in some acidic soils. When toxicity is severe enough, plant growth is retarded, which may result in the soil becoming susceptible to erosion by water and wind and not conducive to water conservation because of limited or no plant cover or plant residues on the soil surface.

### *Antecedent Soil Water Content*

Antecedent soil water content—namely, the degree of soil wetness at the start of an irrigation or a precipitation event—influences the amount of water that will infiltrate the soil from that event. A low antecedent water content may result in little runoff and provide for storing close to the maximum amount of applied water in soil. In contrast, when the antecedent water content is high, runoff usually is high, which results in little water conservation, especially of water from precipitation (irrigations can be delayed to minimize runoff when the antecedent soil water content is high). When runoff occurs, the potential for erosion by water generally is present.

### *Available Water Capacity*

The available water capacity is the capacity a soil has for storing water that can be used by plants. It is the difference between the amount retained at field capacity and the permanent wilting point. The field capacity is the amount of water retained in a soil after it has been saturated and drainage becomes negligible (usually after 2 or 3 days); the permanent wilting point is essentially the lower limit of water availability to plants from the soil. Under laboratory conditions, the field capacity and permanent wilting point are estimated by soil water contents at matric potentials of –0.03 and –1.5 MPa, respectively, but the available water capacity does not necessarily indicate the amount of water plant roots can extract because the amount is plant specific. Soils of different texture differ in their capacity to store water, which directly affects the potential for water conservation. A soil with a low storage capacity may be quickly filled; thus further conservation is not possible. In contrast, the potential for additional water conservation may exist for a relatively long time for a soil with a greater storage capacity.

### *Biological Activity*

Activities by soil microorganisms, soil-inhabiting insects, and so on improve soil permeability to water and thereby reduce the potential for erosion by water and improve conditions for water conservation. These benefits result from holes formed by crickets, earthworms, and the like, which promote more rapid water entry into soil, and by the stabilizing effects of microorganisms in soil.

### *Bulk Density*

Soil bulk density is the mass of dry soil per unit of bulk soil volume. Soils with a high density usually impede water infiltration and root growth, thereby potentially leading to greater soil and water losses.

### *Compaction*

Soil compaction involves the void (pore) spaces between soil particles being decreased, which increases the soil's bulk density. Compaction under crop production conditions results from tractor and implement traffic, trampling by animals, and natural settling of soil. In general, compaction leads to greater runoff, increased potential for erosion, reduced water infiltration, and decreased plant growth.

### *Conditioner*

A conditioner is any substance applied to a soil to improve its physical condition, for example, to reduce its density, stabilize its aggregates, or alter its pH to enhance plant growth and water infiltration. Conditioners include sawdust, compost, peat, synthetic polymers, and various inert materials (*see* **SOIL, *Additive/Amendment***). Improved soil physical conditions generally reduce the potential for erosion and increase the potential for water conservation.

### *Conserving Crops*

In terms of soil and water conservation, soil-conserving crops are those that, for example, adequately cover the surface to shield it against raindrop impact, provide adequate stems per unit area to impede water flow across the surface, have a root system that holds soil particles in place, produce adequate residues to protect the land between crops, and maintain or replenish soil organic matter.

### *Cracks*

Cracks can provide for rapid water entry into a soil, thereby reducing runoff and the potential for erosion and increasing the potential for water conservation. Soil cracking, however, can also result in increased water losses as a result of evaporation.

### *Crusting*

Soil crusting is the result of soil aggregate disintegration and rearrangement of soil particles to form a relatively smooth and dense layer at the surface. Aggregate disintegration may be due to impact by raindrops or sprinkler irrigation or to wetting by flowing water (runoff or irrigation). Crusting may seal the surface, thereby increasing runoff and the potential for erosion and reducing water conservation. Crusting may also impede seedling emergence and reduce soil aeration, thereby resulting in poor plant establishment, growth, and production, which can also hinder soil and water conservation.

### *Degradation*

Soil degradation is the lowering of a soil's suitability for productive purposes. As production decreases, soil degradation is further increased. Degradation may result from erosion by water or wind, accumulation of excess salts, chemical imbalances, and biological alterations. Such degradation may be caused, for example, by using excessive or improper tillage, overgrazing, improper fertility practices, and excessive removal of crop residues.

### *Depleting Crops*

Soil-depleting crops are those that tend to deplete soil nutrients and organic matter, which leads to soil structure deterioration. With continued use of such crops without supplying the necessary nutrients, soil productivity eventually declines, thus increasing the potential for soil and water losses because of poor plant growth.

### *Effective Depth*

The effective depth of a soil is the depth to which plant roots can readily penetrate to obtain water and nutrients. It is the depth to a layer or zone in the soil profile that differs sufficiently in its physical or chemical properties from the overlying material to seriously retard or prevent root growth. To achieve satisfactory plant growth and production and, therefore, to minimize the potential for soil and water losses, effective soil depth must be considered when selecting a crop for a given site. A deep-rooting crop usually does poorly in a shallow soil. A shallow-rooting crop may not fully utilize water and nutrients available in the soil, and, as a result, water and nutrients may move too deep in the profile to be extracted by the plant roots.

### *Fertility*

A proper soil fertility level, which is the ability of a soil to supply adequate nutrients in proper balance to plants, is essential for optimum growth and production of a crop when other growth factors such as water, temperature, light, and physical conditions of the soil are favorable. With an improper level, poor plant growth may result in poor surface cover and low production, which may result in increased potential for soil and water losses.

### *Heaving*

Heaving results from freezing of a wet soil in winter. Heaving partially lifts plants out of the soil, breaks their roots, and causes plant desiccation, which may lead to the surface becoming bare and, therefore, susceptible to erosion and unfavorable for water conservation.

### *Horizon*

A horizon is a layer of soil lying approximately parallel to the land surface and differing from adjacent layers in physical, chemical, and biological properties or characteristics such as texture, structure, consistency, color, and degree of acidity or alkalinity. Horizons at a given site constitute the soil profile for that site (*see* **SOIL, *Profile***). The influence of a horizon on soil and water conservation is mainly through its physical characteristics. Characteristics of the upper horizon strongly influence the potential for erosion and water conservation. Water conservation, especially the capacity of the soil to store water, is also influenced by characteristics of deeper horizons.

### *Humus*

Humus is the total of organic materials or compounds in a soil. It does not include undecayed or partially decomposed plant and animal tissues in the soil biomass. Soil humus content influences soil structure, thereby also influencing soil water relations (e.g., infiltration and water-holding capacity) and plant rooting, growth, and production, which, in turn, influence the potential for soil and water conservation (*see* **SOIL, *Organic Matter***).

### *Imprinting/Patterned Soil*

Soil imprinting uses a massive steel roller that has, for example, two patterns of angular steel teeth that form relatively stable impressions (patterned imprints) as the roller is pulled across the land surface. The practice was developed to improve vegetation on overgrazed and shrub-infested rangelands in arid to semiarid regions while protecting the land against accelerated runoff and erosion. During its operation, the roller crushes the shrubs and other surface materials and presses them into the soil surface, thereby helping to stabilize the imprints. This reduces the potential for surface sealing, which would tend to increase runoff. Grass can be seeded while the operation is performed. When it rains, the action of raindrops and flowing surface water moves seed, topsoil, and plant litter into the surface imprints, where they are concentrated along with the water to enhance the probability of successful germination and seedling establishment. Of course, the prime requisite for successful vegetation establishment in such arid and semiarid regions is adequate precipitation.

### *Inherent Productivity*

A soil's inherent productivity is its productivity status when it is first developed for crop production from native conditions. Under some conditions, as in the Great Plains of the United States, the inherent productivity was relatively high for the crops initially grown but has declined substantially. As a result, present crop production requires increased inputs to obtain adequate plant growth to provide conditions conducive to soil and water conservation. In other cases, for example, land developed from the sea by building dikes and pumping water from behind the dikes, major inputs are needed to achieve a satisfactory productivity level and, thereby, to achieve soil and water conservation.

### *Inversion*

The surface layer of some soils consists of materials (sands) that are highly erodible, have a low water-holding capacity, or have some other undesirable characteristic. When a layer consisting of a more desirable material lies below this layer, inverting the layers by, for example, deep plowing can bring the more desirable material to the surface. Such inversion of the layers can improve water storage in soil, reduce the potential for erosion, and, thereby, improve conditions for crop production. Through improved crop production, the potential for subsequent soil and water losses is reduced.

### *Liming/Soil pH*

Liming is the process of applying lime to a soil to reduce soil acidity and supply calcium for plant growth (dolomitic limestone also provides magnesium). Benefits of liming in terms of improving soil and water conservation result from improving the growth of plants that are not adaptable to acidic soils (*see* **SOIL,** ***Additive/Amendment,*** *Lime/pH*).

### *Limiting Horizon*

A limiting horizon in a soil profile is one that limits or prevents water movement, aeration, and/or root growth at some depth in the soil. Soil and water conservation may be reduced when a limiting horizon is present because of limited plant growth, with the degree of reduction being strongly influenced in most cases by the depth of the limiting horizon.

### *Loosening*

A surface crust or dense layer near the soil surface generally results in increased runoff, which can increase erosion by water and decrease water conservation. Erosion by wind may occur under some conditions when a surface crust is present. A crust can also retard or prevent seedling emergence, thereby resulting in poor crop establishment, untimely establishment of plant cover, and poor crop yields. Tillage at a shallow depth is satisfactory for loosening the crust and generally improves conditions for water infiltration, which reduces the potential for erosion by water and/or improves water conservation. The potential for erosion by wind is reduced if tillage that loosens the crust also adequately roughens the soil surface. In addition to increasing runoff, a dense subsurface layer may also retard or prevent root penetration, thereby reducing crop productivity. Loosening of a subsurface layer is accomplished with a chisel or similar implement. This improves conditions for reducing runoff, reducing erosion by water, increasing soil water storage, and enhancing plant rooting, thereby improving conditions for crop production.

### *Organic Matter*

Soil organic matter is the organic fraction in a soil that consists of plant and animal residues in various stages of decomposition, cells and tissues of soil organisms, and substances produced in the soil by plant roots (exudates) and other organisms. Soil organic matter content strongly influences soil structure and, therefore, soil water relations (e.g., infiltration and water-holding capacity), plant rooting, plant growth, and crop productivity, which affect soil and water conservation (*see* **SOIL, *Humus***).

### *Patterned Soil*

*See* **SOIL, *Imprinting/Patterned Soil.***

### *Permeability*

Permeability relates to the ability of water, air, and plant roots to penetrate or pass through the bulk mass or specific layers of a soil. Soil permeability, therefore, affects water runoff, infiltration, and conservation and plant rooting, growth, and production, which, in turn, may directly or indirectly affect the potential for erosion by water and wind.

### *pH*

Soil pH strongly affects which plants can be grown in a soil. Although there are plants that grow well in low-pH soils and others that grow well in high-pH soils, extremely low and high pH values generally result in poor or no growth of plants, which may increase the potential for erosion by water and wind and decrease the potential for water conservation (*see* **SOIL, *Additive/Amendment,*** *Lime/pH*).

### *Plow Pan*

A plow pan is a dense, compacted horizon or layer formed at the depth of tillage by repeatedly tilling a soil at the same depth with certain implements. A plow pan can retard or prevent water movement and plant root growth into the soil, thereby increasing the potential for runoff, soil and water losses, and low crop productivity.

### *Polymer*

*See* **SOIL, *Additive/Amendment,*** *Polyacrylamide.*

### *Porosity*

Soil porosity is that part of the total soil volume not occupied by solid particles. It is the space between the solid particles and, therefore, strongly affects water, air, and root penetration into the soil. These, in turn, affect water runoff, infiltration, and conservation; plant rooting, growth, and production; and the potential for erosion (*see* **SOIL, *Permeability***).

### *Previous Erosion*

Previous erosion by water or wind can strongly affect subsequent soil erosion. For example, if water flow from a rainstorm has caused the formation of a rill or gully, a subsequent rainstorm will cause water to again flow in the same rill or gully and enlarge it, unless the soil surface has been altered between the storms. Likewise, if erosion by wind has occurred on an area or damaged or destroyed plants on the area during a storm, the next storm most likely will again cause soil loss from the same area unless erosion control measures have been applied.

### *Profile*

A soil profile is a vertical section of a soil from its surface through all of its horizons, including the C horizon, which is the lowest horizon in a soil profile. The C horizon consists of materials similar to the underlying materials from which the soil developed and is little affected by activities in the overlying horizons of the profile. A profile influences soil and water conservation through the characteristics of the individual horizons that constitute the profile (*see* **SOIL, *Horizon***).

### *Profile Modification*

Soil profile modification is the process of altering a soil profile by deep plowing or using other equipment to such a depth that the materials from two or more horizons become homogenized. Such modification results in improved water infiltration, plant root growth, and crop yields on an otherwise slowly permeable soil. Through improved plant growth and production, the potential for soil and water losses is decreased.

### *Rocks/Stones*

*See* **SOIL**, ***Stoniness.***

### *Saturated Soil*

A saturated soil is one that has all pores filled with water. For most saturated soils, except those with a high sand content, water infiltration is low, which increases runoff and, therefore, the potential for erosion by water. Also, further water conservation is impossible because no additional storage space is available. Soil saturation prevents aeration, which is detrimental to the growth and productivity of many plants, and, therefore, could lead to erosion by water under some conditions.

### *Self-Mulching Soil*

A self-mulching soil is one whose surface layer consists of highly water-stable aggregates that do not disintegrate as a result of raindrop impact. Thus, surface sealing and crusting do not occur and the aggregates serve as mulch when drying occurs. The potential for runoff and erosion by water generally is low and water conservation generally is favorable on such soils. During intense storms, however, runoff may occur, which may cause the relatively loose aggregates to be moved by the flowing water.

### *Shrinking/Swelling*

Soil shrinking and swelling result from the drying and wetting of soils that contain certain types of clay (e.g., montmorillonite). When the soils dry, cracks develop, which usually provide for rapid water entry into such soils and a high potential for water conservation when rain occurs or irrigation water is applied. After they are wetted, however, water entry into such soils may become extremely slow, thus increasing runoff and the potential for erosion by water and reducing the potential for storing additional water in the soil.

### *Stabilizer*

A stabilizer is a chemical or mechanical treatment used to increase or maintain the stability of a soil or to improve its engineering properties, thereby reducing the potential for soil and/or water losses. Such practices include applying emulsions (*see* **SOIL**, Additive/Amendment, *Emulsion*), land imprinting (*see* **SOIL**, ***Imprinting/Patterned Soil***), and applying mulch to protect an earthen structure, roadside, or waterway until a permanent vegetative cover becomes established.

### *Stoniness*

Soil stoniness refers to the proportion of stones (or rocks) in a soil or on its surface. It is a characteristic used in classifying a soil. Stones often are removed from soil to improve the performance of cultural operations. Stones left in place may provide for soil and water conservation by protecting the surface against raindrop impact and retarding water flow across the surface.

### *Structure*

The structure of a soil refers to the combination or arrangement of primary particles into secondary units called peds. Peds are classified according to their size, shape, cohesion, degree of distinctness, and stability. Recognized soil structures are structureless (no observable formation of peds), weak, moderate, and strong (distinct peds in an undisturbed soil). A soil with a strong structure usually is more conducive to soil and water conservation than one with a weak structure, but water infiltration may be satisfactory on a soil made up of a deep sand that is considered to be structureless. However, although infiltration may be satisfactory, relatively little water may be conserved because such sandy soils have a low storage capacity. On such soils, deep percolation (or drainage) may be high and the water may replenish an aquifer and contribute to water flow in a spring or stream. Another structureless condition is massive, for which the entire soil mass is strongly coherent. As a result, erosion by water and wind usually is low because there are no definite lines of weakness in the soil mass.

### *Subsidence*

Soil subsidence, which is a type of mass movement, generally results in very little soil removal. When the subsidence is relatively deep, some soil sliding and washing may occur, which is a type of erosion by water. Where subsidence is severe, the surface may be too severely broken to be suitable for some agricultural uses.

### *Subsurface Barrier*

A subsurface barrier is a layer of some material placed in a highly permeable soil (e.g., a deep sand) to minimize deep movement of water, thereby retaining the water within reach of plant roots. Plastic films and asphalt layers have been used as subsurface barriers to reduce water losses from the root zone of plants, thus resulting in water conservation.

### *Surface*

Conditions at the soil surface strongly influence the potential for erosion and water conservation. The surface is the point where erosion occurs and where water must enter the soil to be conserved. Numerous surface conditions affect soil and water conservation, and many are discussed under other topics. The following are additional conditions or terms that pertain to the soil surface.

#### *Cloddiness*

A cloddy soil surface usually is effective for controlling erosion by wind, provided adequate clods of sufficient size are present and the soil material resists the forces of the wind.

### *Configuration*

*See* **WATER CONTROL PRACTICES, *Microwatershed.***

### *Modification*

Modification of a soil surface includes the use of such practices as tillage, terracing, mulching, and furrow diking.

### *Physical Characteristics*

Physical characteristics strongly affecting soil and water conservation include texture, aggregation (aggregate stability), bulk density, and compaction.

### *Ridges*

Ridges result from the use of various tillage or crop planting implements. Ridges on the contour (across the slope) of the land help control runoff, thereby helping to control erosion by water and improving water conservation by retaining water on the land. Ridges aligned perpendicular to prevailing winds help control erosion by wind.

### *Roughness*

Surface roughness can be achieved by tillage that leaves the surface in a rough and/or cloddy condition or creates ridges and furrows on the land. Such roughness may provide for temporary water storage in depressions (detention storage) on the surface or in large pores near the soil surface, thereby providing more time for water infiltration and storage in soil. Such temporary storage also reduces the potential for erosion. Roughness resulting from surface clods and ridges helps control erosion by wind.

### *Sealing*

Surface sealing occurs when soil aggregates are dispersed as a result of raindrop impact or when wetted by irrigation water, with the soil particles then moved into soil pores. Surface seals consist of a thin soil layer but greatly reduce water infiltration, thereby increasing runoff and erosion by water and minimizing the potential for water conservation.

### *Slope*

In general, the potential for runoff and erosion by water increases and for water conservation decreases with increases in surface slope. As a result, the need for soil and water conservation practices increases with slope increases.

### *Stabilization*

Numerous practices are effective for stabilizing a soil surface, including mulches, emulsions, and mechanical treatments (tillage or imprinting).

## ***Survey/Land Classification***

A thorough knowledge of prevailing soil and land conditions is essential for developing effective soil and water conservation practices for a given tract of land because not all available practices are equally effective for use under all conditions. Unfortunately, many soil surveys are outdated, incorrect, or incomplete. In other cases, soil surveys are not available. Likewise, land classifications often are not complete or lack sufficient detail to provide the information needed to develop effective conservation plans for a given tract of land. Under such conditions, development and implementation of effective conservation measures usually are limited or do not occur. Surveys should include enough detail to show the different soils that occur on the tract of land under consideration. Different soil series often occur close together, with each series possibly requiring different practices for greatest success in achieving soil and water conservation.

### *Texture*

Soil texture pertains to the relative proportions of the different soil separates (sand, silt, and clay) in the soil mass. The soil textural classes are clay, clay loam, loam, loamy sand, sand, sandy clay, sandy clay loam, sandy loam, silt, silty clay, silty clay loam, and silt loam. Soil textures, in general, have a strong influence on soil and water conservation. Water infiltration generally is high for coarse-textured (high-sand-content) soils, and relatively high water runoff rates are needed to transport individual soil particles. Such soils, however, usually have a single-grain structure and, therefore, may be highly erodible by water and wind when not adequately protected. Soils with a relatively high clay content generally develop aggregates that provide good resistance to dispersion when wetted, thus minimizing surface sealing that reduces infiltration and increases runoff and the potential for erosion by water. Some soils with a high silt content are highly susceptible to erosion, especially when the silt particles are not associated with enough clay to produce a well-aggregated or granular soil structure. Surface sealing is a major problem when such soils are wetted, which reduces infiltration, increases the potential for erosion, and minimizes the potential for water conservation.

### *Topsoil Removal*

Erosion by water and wind removes topsoil and plant nutrients from the land. Unless controlled, the erosion eventually exposes the subsoil, thereby generally resulting in lower plant growth and productivity. Because of reduced plant growth, further erosion may occur, thereby accelerating land degradation and making it increasingly difficult to control erosion and achieve satisfactory plant production.

### *Variability*

Soil variability, for example, land slope, soil texture, and fertility level, may differentially influence crop performance within a management unit (field). Because of differential plant growth, the potential for erosion by water and wind and for water conservation may differ at different sites within the unit. Different practices may thus be required at different sites in the field for most effective crop performance and for soil and water conservation (*see* **AGRICULTURE, *Precision Agriculture***).

### *Water Content*

*See* SOIL, ***Antecedent Soil Water Content.***

## SOILS

In addition to the characteristics or conditions of a particular soil, as discussed under **SOIL,** soils are classified according to combinations of specific characteristics. The following have a major effect on the potential for soil and water conservation.

### *Dense Soils*

Water infiltration generally is low for soils with a naturally dense subsurface horizon. As a result, runoff usually is high, which increases the potential for erosion by water and limits the potential for water conservation. The dense horizon in such soils also greatly reduces or prevents root penetration, thereby possibly resulting in inadequate plant growth for protection against soil and water losses.

#### *Duripan*

A duripan is a subsurface horizon cemented by silica that is highly stable in water.

#### *Fragipan*

A fragipan is a subsurface horizon with a bulk density much greater than the soil material above it. It appears to be cemented when dry but is moderately to weakly brittle when moist. The horizon has a low organic matter content and is slowly or very slowly permeable to water.

### *Petrocalcic Horizon*

A petrocalcic horizon is a continuous indurated calcic horizon that is cemented by calcium carbonate and, at some places, by magnesium carbonate. A petrocalcic horizon prevents penetration by plant roots.

### *Petrogypsic Horizon*

A petrogypsic horizon is a continuous, massive gypsic horizon that is strongly cemented by calcium sulfate. A petrogypsic horizon prevents penetration by plant roots.

### *Placid Horizon*

A placid horizon is a thin (1 to 25 mm) mineral horizon that is commonly cemented by iron. A placid horizon is slowly permeable or impermeable to water and plant roots.

## *Hydrophobic Soils*

Hydrophobic soils are water repellent (difficult to wet with water). This water repellency often is due to dense fungal mats or hydrophobic substances vaporized and redeposited during a fire. Forested lands often are highly erosive after a fire.

## *Organic Soils*

Organic soils contain a high percentage (generally greater than 20 or 30 percent) of organic matter throughout the profile. Organic soils may be erodible by water. They are not erodible by wind when wet but usually are highly erodible by wind when they become dry and when there is limited or no surface protection.

## *Peat Soils*

Peat soils are organic soils in which plant residues can be recognized; that is, the residues have not decomposed or are only slightly decomposed. The organic matter content is greater than 50 percent. Peat soils occur primarily in humid regions, and erosion usually is not a problem. However, when precipitation becomes sparse and the surface is devoid of vegetation, erosion by wind can be severe.

## *Saline-Alkali Soils*

Saline-alkali soils contain sufficient exchangeable sodium to adversely affect the growth of most plants. They contain appreciable amounts of soluble salts or have a combination of harmful qualities of salts and either high alkalinity or a high amount of exchangeable sodium (or both) distributed in the soil profile. Poor plant growth (or failure to grow) under such conditions may leave the soils highly susceptible to erosion by water and wind. Water conservation generally is low because of poor infiltration on these soils.

## *Saline-Sodic Soils*

Saline-sodic soils contain sufficient exchangeable sodium to adversely affect growth of most plants. Such soils also contain appreciable quantities of soluble salts. Poor plant growth (or failure to grow) under such conditions may leave the soils highly susceptible to erosion by water and wind. Water conservation generally is low because of poor infiltration on these soils.

## *Saline Soils*

Saline soils are nonsodic soils that contain sufficient soluble salts to be detrimental to crop growth and productivity but do not contain an excessive amount of exchangeable sodium. Poor plant growth (or failure to grow) under such conditions may leave the soils highly susceptible to erosion by water and wind. Water conservation generally is low because of poor infiltration on these soils.

### *Salt-Affected Soils*

Salt-affected soils are those on which the presence of soluble salts, exchangeable sodium, or both adversely affects the growth of most crop plants. Poor plant growth (or failure to grow) under such conditions may leave the soils highly susceptible to erosion by water and wind. Water conservation generally is low because of poor infiltration on these soils.

### *Sodic Soils*

Sodic soils are nonsaline soils containing adequate amounts of exchangeable sodium to adversely affect the soils' structure and crop productivity under most conditions. The poor crop growth may increase the potential for erosion by water and/or wind and the poor structure may result in limited water conservation.

## STRAW SPREADER

A straw spreader is an attachment on a grain harvester that spreads the nongrain plant material that passes through the harvester relatively uniformly over the entire area from which it is cut. Without such an attachment, the material is usually deposited in a narrow swath directly behind the harvester. Spreading the straw provides a relatively uniform cover on the soil surface, which is desirable for soil and water conservation purposes.

## STREAM EROSION CONTROL

The banks and channels of many streams are susceptible to erosion (e.g., "washing away" or collapse), especially when the flow rate and volume are high, as during floods. Such erosion increases the sediment load in streams, which eventually increases the amount deposited in reservoirs or at other downstream sites. Streambank erosion also decreases the amount of land available for other uses and causes unsightly conditions.

### *Bank Stabilization*

Streambanks can be stabilized by planting trees and other plants on the banks and by placing nonerosive materials (e.g., riprap and brush matting) at critical locations on the streambanks. The potential for streambank erosion can be decreased also by implementing channel improvement or realignment measures such as clearing, widening, deepening, or straightening the existing channel or even developing a new channel.

### *Bendway Weir*

The bendway-weir system of streambank erosion control consists of a series of rock jetties (weirs) built along an eroding streambank. The jetties start at the streambank, are oriented upstream at an angle of 15° to 20°, and extend to about half the width of the stream at low-water levels. A series of four to six such weirs cause water to be directed toward the middle of the stream, with sediments being deposited at the downstream side of the weirs, where water flow rates are lower. At greater flow rates, water flows over the weirs.

### *Bolster (Rock)*

A rock bolster is a loose rock-filled dam made of readily available stones held in place by wire netting. The dam is constructed by placing heavy-gauge galvanized wire netting across the gully, filling half of the wire with stones, folding the other half over the stones, and lacing the edges together, thereby forming a sausagelike dam or bolster of rocks. Such a dam slows the flow of water, thereby providing temporary protection while vegetation is established in the gully.

### *Box Inlet*

A box inlet is a structure placed in a waterway or drainageway to allow runoff or drainage water to be conveyed nonerosively to a lower elevation on relatively steeply sloping land. Each structure provides for a drop in elevation, thus creating a gentler slope where the water flows downhill between successive downslope structures.

### *Gabion*

Vegetative barriers (trees, grasses, etc.) sometimes do not satisfactorily control streambank erosion because of rapid flow rates and steep banks. Under such conditions, rigid structures also may not be satisfactory because of their high cost and lack of flexibility. To overcome those problems, gabions are sometimes used. Gabions are baskets, sacks, or similarly shaped containers constructed of heavy-gauge wire netting and filled with stones. When anchored in the desired location, they help control erosion. Similar wire-enclosed stone structures are used to control erosion on roadside embankments.

### *Graded Streambank*

A graded streambank is one shaped to an appropriate slope to minimize the potential for further erosion as a result of runoff into the stream or flowing water in the stream. Establishing vegetation (e.g., trees or grass) or placing brush matting or stones (*see* **STREAM EROSION CONTROL, *Gabion***) on the graded bank further reduces the potential for subsequent erosion.

### *Grass-Lined Channel*

A grass-lined channel (or waterway) is a channel designed to convey excess water from land at nonerosive flow rates. Such channels often are used in conjunction with terraces or drainage channels.

### *Groin*

A groin (or jetty) is a structure (wall) built out into the water to deflect a current away from a streambank, pier, or harbor, for example, thereby reducing the potential for erosion caused by flowing water or wave action. It is constructed either at a right angle to the bank or with a downstream deflection. Groins can be built with vertical timber pilings, concrete, or stones.

### *Jetty*

*See* **STREAM EROSION CONTROL, *Groin*.**

### *Pool and Riffle Method*

The pool and riffle method consists of a series of rock and gravel bars (riffles) constructed across a stream at a spacing of approximately six times the width of the stream. Such riffles break a relatively steeply sloping section of a stream into more gently sloping sections, with similar results to those achieved with terraces on a steeply sloping eroding hillside.

### *Reno Mattress*

A Reno mattress is a wire mesh basket about 2 m wide and 6 m long that is filled with rocks and stones to a depth of ~0.2 to 0.3 m. Each basket is divided into several sections lengthwise to prevent the rocks or stones slipping to the lower end. The baskets are placed side by side to protect sloping streambanks against erosion caused by flowing water (*see* **STREAM EROSION CONTROL, *Gabion***).

### *Revetment, Green-Tree*

A green-tree revetment is an overlapping layer of freshly cut trees, with limbs and branches intact, placed along streambanks to control erosion. Strong steel cables anchored to logs ~6 or 7 m apart and buried 2 to 3 m deep in the remaining streambank hold the trees in place.

When water volume in the stream increases and water flows over the trees, the streambank rebuilds itself with sediments deposited in voids between the branches.

### *Riprap*

Riprap is a sufficiently thick layer of rocks, stones, or boulders of suitable size to resist the erosive forces of wave action or flowing water. Riprap usually is used to protect shores, the slopes of earthen dams, and the outlets of water control structures and in channels where flow rates are relatively high.

### *Roots, Matted Tree*

The matted root systems of trees such as willow or poplar are highly effective for stabilizing streambanks, thereby reducing the potential for flowing water to cause streambank erosion (*see* **STREAM EROSION CONTROL, *Willow Post***).

### *Sod Check*

Sod checks are sod strips or sodded earth fills placed across a gully to assist the establishment of vegetation in the gully and to stabilize channels until vegetation between the checks becomes established. The strips or fills are ~0.3 m wide and are placed flush with or slightly below the bed of the gully at ~2 m intervals.

### *Sod Chute*

A sod chute is a waterway or spillway in which a permanent grass has been established to minimize the potential for erosion when water flows within it from one elevation to a lower elevation (*see* **WATER CONTROL PRACTICES, *Waterway***).

### *Stone Toe*

The stone-toe system consists of rocks piled along the eroding bank of a stream. To avoid undercutting of the bank, additional linear piles of rocks are placed behind and perpendicular to the rocks at the bank.

### Stream Barbs

Stream barbs are similar in design to bendway weirs. Stream barbs, however, are constructed of larger rocks and point upstream at an angle of 70°. Because larger rocks are used for the barbs, the number needed is about half that for bendway weirs under the same stream conditions.

### Willow Post

The willow-post method consists of installing (planting) dormant willow cuttings (posts) in the streambank. The posts should be ~7 to 10 cm in diameter and ~3 to 5 m long. They are placed in holes drilled in the streambank to a sufficient depth (1.5 to 2 m) to be securely anchored in the soil. Three to fives rows of posts are used. The first row should be in the stream, with successive rows up the streambank at ~1.2 m intervals. Bushes of other trees can be placed between the two lowest rows of trees with their cut end pointing upstream. These are used to protect the streambank while the freshly planted willow posts develop their root systems. They also help trap sediments. The willow posts develop a dense root system that is highly effective for controlling streambank erosion.

## STRIPPER HEADER

A stripper header on a harvester removes the grain and a small portion of other materials (mainly the chaff) during the harvesting of small grains (e.g., wheat). Harvesting with such equipment results in most plants remaining standing, with limited amounts of fine materials on the soil surface and no concentrations of straw behind the harvester. This differs from harvesting using a cutter header that cuts the plants below the heads and deposits the straw in a swath behind the unit (unless the harvester is equipped with a spreader). The straw can also be spread in a separate operation when harvesting is done with a cutter header. Using a stripper header decreases the residue decomposition rate and residue spreading is not needed. In addition, the standing residues promote greater shielding of the soil surface against wind and radiant energy, which is conducive to reduced evaporation and, therefore, improved water conservation.

## SUSTAINABLE AGRICULTURE/ SUSTAINABILITY

Sustainable agriculture aims at managing and conserving the resource base to ensure its availability and productivity for continued use by present and future generations. It strives to conserve soil, water, plant, and animal resources; minimizes or avoids environmental degradation; and is technically appropriate, economically viable, and socially acceptable. Management practices that control erosion by water and wind and provide for water conservation where needed ensure the sustained productivity of natural resources.

# T

## THRESHOLD VALUE

The threshold value is the wind velocity required at the soil surface to initiate the detachment and movement of soil particles, which results in erosion by wind.

## TILLAGE

Most crop production systems involve using one or more types of tillage, the mechanical manipulation of a soil relatively near its surface, during the crop cycle. Tillage is used to prepare seedbeds and root beds, control weeds, aerate the soil, disrupt dense soil layers, incorporate plant nutrients, plow down crop residues, and control erosion. Most tillage operations are performed before crop establishment, but later tillage (*see* **TILLAGE, *Cultivation/Cultivating***) often is used to control weeds and aerate the soil. Tillage can be used for rangeland and pasture improvement and is also used to manipulate the soil at construction sites (highways, roads, building sites, etc.). The type of tillage used has a strong influence on the potential for erosion by water and wind and for water conservation in many cases.

### *Basin Listing*

*See* **TILLAGE, *Furrow Diking.***

© 2006 by The Haworth Press, Inc. All rights reserved.
doi:10.1300/5678_17

### *Blade Plow*

A blade plow is an implement with shanks (or legs) on which a blade is mounted. As the implement is drawn through the field, the blade undercuts the surface to sever the roots of weeds and loosen the soil to improve the seedbed but retain crop residues on the surface for erosion control and water conservation purposes. Blade tillage is widely used to control erosion by wind and it helps control erosion by water and improves water conservation in many cases (*see* **TILLAGE, *Stubble Mulch Tillage; Subsurface Tillage***).

### *Chain*

A chain tillage implement consists of a section of a ship's anchor chain to which specially shaped blades have been welded. The chain is mounted between two points on a frame with a mechanism that allows the chain to rotate as it is drawn across the land. Used on a relatively loose or previously loosened soil, the heavy implement makes depressions ~10 cm deep in the surface, providing for water to be captured to improve grass establishment. This practice also reduces the potential for runoff, thus reducing the potential for erosion by water.

### *Chemical Fallow*

Chemical fallow is the practice of controlling or killing all weeds during the fallow period with chemicals (herbicides). Tillage may or may not be used with this practice. Chemical fallow results in crop residues being retained on the surface, which is highly effective for controlling erosion and improves conditions for water conservation.

### *Chisel*

A chisel has gangs of closely spaced narrow shanks for loosening a soil or disrupting a dense soil layer at a greater depth (generally 20 to 40 cm) than is achieved with normal tillage (usually less than 20 cm). This generally improves water infiltration and plant rooting depth and growth, which enhances the potential for erosion control and water conservation.

### *Chisel-Chopper*

A chisel-chopper is a specialized implement with chisel shanks to loosen the soil and a rolling chopper to cut crop residues and weeds. It loosens the soil but retains most plant materials on the surface for erosion control and water conservation purposes.

### *Clean Tillage/Clean Culture*

Clean tillage is the practice of plowing and cultivating to incorporate all crop residues and control all vegetation, except for the crop being grown. It can result in a high potential for erosion by water and wind and poor conditions for water conservation in some cases. On relatively flat soils (low surface slope) and with adequate surface roughness or depressions, clean tillage may not cause erosion or water conservation problems.

### *Conservation Tillage*

Conservation tillage is any combination of tillage and planting practices that generally reduces the loss of soil and water relative to losses with conventional tillage (*see* **TILLAGE, *Conventional Tillage***). The term covers any tillage method that retains protective amounts of crop residues on the soil surface. In the United States, any method that provides for a 30 percent or greater cover of crop residues after planting is considered to be a conservation tillage method.

### *Conventional Tillage*

Conventional tillage is the combination of primary and secondary tillage operations normally used for seedbed preparation and growing-season weed control for a given crop in a given region. Conventional tillage usually results in less than a 30 percent cover after crop establishment, which may increase the potential for soil and water losses under some conditions compared with losses from the use of conservation tillage (*see* **TILLAGE, *Conservation Tillage***).

### *Cultipacker*

A cultipacker has wide rollers with corrugated or jagged working surfaces and is used for broadcast crushing or firming a soil, for example after tillage with a moldboard plow. The roughened or ridged surface that results may provide for temporary water storage, thereby improving control of erosion by water and enhancing water conservation. Control of erosion by wind may also improve if the cultipacker adequately stabilizes the soil surface.

### *Cultivation/Cultivating*

Cultivation is the practice of using shallow tillage to control weeds or to provide improved conditions for soil aeration and water infiltration and storage, all of which are important for obtaining good plant growth and crop productivity and thus improved conditions for erosion control. Good crop production generally provides for greater income for the producer or landowner, thus making further use of conservation practices more likely.

### *Dammer-Diker*

A dammer-diker is an implement that makes depressions in the soil with paddle-like blades on a rotating shaft as the implement is drawn across a field. The depressions retain water from precipitation on the surface, thereby providing more time for infiltration and increasing the potential for water conservation.

### *Deep Plowing*

Deep plowing refers to any plowing to a depth greater than normal tillage, which is usually less than 20 cm. It usually is done to loosen a dense or compacted zone in the profile; sometimes it is done to invert soil zones, either to bring more desirable soil materials to the surface or to bury some undesirable soil material or other substance present at the surface (*see* **SOIL, *Inversion***). Deep plowing that loosens a dense soil zone usually improves water infiltration, thereby reducing runoff and erosion by water, improving water conservation, and improving root growth. Deep tillage that brings clods to the surface helps control erosion by wind.

### *Deep Ripping*

*See* **TILLAGE, *Subsoiling.***

### *Diker (Paddle)*

A diker is used to form small earthen dikes (or blocks) in furrows (*see* **TILLAGE, *Furrow Diking***). A diker may be a simple tool such as a spade used by hand or a mechanically or hydraulically operated implement powered by a tractor.

### *Direct Drilling*

*See* **TILLAGE, *No-Tillage.***

### *Disk Implements*

Disk implements have concave disks to cut into and turn the soil as the implement is pulled through the field. Disk tillage is performed for such purposes as controlling weeds; incorporating crop residues, plant nutrients, and pesticides; loosening the soil; and, under some conditions, roughening the soil surface to help control erosion by wind by bringing clods to the surface. When properly adjusted, most disk implements leave the surface relatively even, with a certain level of roughness that depends on soil texture and soil water content when the operation is performed. Such roughness can provide for temporary water storage on the surface, thereby reducing runoff and the potential for erosion by water and improving water conservation. When limited amounts of residues are available, incorporating the residues may increase the potential for erosion and reduce the potential for water conservation. Another disadvantage of using disk implements is that they cause compaction at the depth to which the disks penetrate the soil under some conditions, thereby reducing water infiltration, increasing runoff and erosion by water, and reducing water conservation. Large disk implements are used to pulverize soil at highway and building construction sites, allowing the soil to be more easily compacted. Similar results under agricultural conditions reduce the potential for soil and water conservation.

### *Disk Bedder*

A disk bedder is used to form beds or ridges similar to those formed with a lister (*see* **TILLAGE, *Lister***). A gang of disks on each side moves the soil toward the bed or ridge.

### *Disk Harrow*

The disk harrow has curved disks and is widely used to control weeds and to partially incorporate crop residues into soil. Types include the offset-disk harrow, which has one front gang that moves soil in one direction and one rear gang that moves soil in the opposite direction, and the tandem-disk harrow, which has two front gangs and two rear gangs, with the front gangs moving soil outward in opposite directions from the center and the rear gangs moving soil inward in opposite directions from the outer edges.

### *Disk Chain*

A disk chain consists of disks attached to a ship's anchor chain. The disk chain rotates as it is pulled across the land. It has been successfully used to improve grass establishment on rangeland and where remnants of brushy trees remain on the land after brush clearing.

### *Disk Plow*

A disk plow results in soil inversion similar to that with a moldboard plow. The soil is mostly inverted to the depth of plowing and relatively little mixing of soil materials takes place. Similar to moldboard plowing, disk plowing is used to loosen the soil and to plow under crop residues or other materials on the soil surface.

### *One-Way Disk*

A one-way disk implement consists of a gang of disks all on one axle, with the axle mounted at an angle on the frame so that operation of the implement moves the soil in one direction. Each tillage operation partially buries some crop residues, and when the one-way disk is used two to four times to control weeds during a period between crops, most residues become buried. This often results in conditions conducive to erosion by water or wind and limited water conservation.

### ***Duck-Foot Cultivator (Field Cultivator)***

A duck-foot cultivator has horizontally spreading, V-shaped sweeps or blades normally used for shallow tillage (5 to 10 cm depth) to control weeds and prepare seedbeds, but it does not invert the surface soil layer or bury crop residues. This retention of surface residues provides conditions conducive to soil and water conservation.

### ***Duck-Foot Sweep***

A duck-foot sweep is one of the V-shaped sweeps or blades on a duck-foot cultivator. It loosens or stirs the soil, thereby roughening the surface and bringing up clods without pulverizing the soil. Implements with duck-foot sweeps are widely used in some semiarid regions (e.g., the Great Plains of the United States) to help control erosion by wind.

### ***Early Implements***

Plowing has been used for centuries to improve crop production and control erosion. A special digging implement, the "mischum," which served much the same purpose as a subsoiler, was used in Greece at the time of Homer. In Scotland, a "shifting moldboard" plow was used on steeply sloping land to retard runoff and reduce erosion in the 1700s and early 1800s. Deep plowing also was met with universal approval at that time because it provided conditions comparable to those achieved with hoe-and-spade cultivation, which produced excellent results on small garden patches in earlier times. Another early implement was an "ard," which remains important in some Mediterranean and Near East countries. This plow has no wheels and no moldboard. As a result, the soil is only loosened and not turned to form a typical furrow. An ard is limited to tillage of lighter soils and should help create a roughened surface with surface depressions, thereby providing conditions for controlling erosion by water and wind and for conserving water.

### *Ecofallow*

*See* CROPPING SYSTEM/SEQUENCE, ***Ecofallow***; *TILLAGE*, ***Chemical Fallow***.

### *Emergency Tillage*

Emergency tillage is used, for example, to quickly roughen the soil surface when erosion by wind is occurring. It often is required when rain has smoothed the surface where clean tillage was carried out and the crop has not yet become established. Emergency tillage is performed with implements such as a sand fighter, rotary hoe, or chisel plow, with the latter operated only deep enough to produce a rough, cloddy surface.

### *Field Cultivator*

*See* TILLAGE, ***Duck-Foot Cultivator (Field Cultivator).***

### *Furrow Diking*

Furrow diking is also known as furrow blocking or tied ridging. The dikes are formed at relatively close intervals to trap water from precipitation or sprinkler irrigations. By retaining water on the land, runoff and erosion are avoided and more time is provided for infiltration to occur, which improves water conservation. During major rainstorms, the dikes may overtop, which may result in erosion by water.

### *Hardpan Disruption*

Tillage to disrupt a hardpan usually is carried out using a chisel plow or a subsoiler, depending on the depth at which the hardpan is located. Disrupting a hardpan improves water infiltration, which reduces the potential for erosion by water and improves conditions for water conservation. It also promotes plant root growth.

### *Harrow*

A harrow usually is an implement that pulverizes, smoothes, or firms a soil to prepare a seedbed, control weeds, or incorporate materials spread on the surface, but some types can be used to loosen the surface soil. A harrow may be a disk, spike-tooth, or spring-tooth implement. Harrowing is a broadcast operation usually considered to be secondary tillage.

#### *Disk Harrow*

*See* **TILLAGE, *Disk Implements,*** *Disk Harrow.*

#### *Spike-Tooth Harrow*

A spike-tooth harrow has rigid teeth that generally smooth and firm the surface, but it can be used to break a surface crust to improve seedling emergence.

#### *Spring-Tooth Harrow*

A spring-tooth harrow has curved, nonrigid teeth that tend to shatter the soil and can be used to roughen the surface when tillage is needed, for example, to control erosion by wind.

### *In-Row Subsoiling*

In-row subsoiling is the practice of using a subsoiler as an integral part of the crop planting equipment to shatter soil ahead of and directly beneath the planted row to improve conditions for plant root proliferation and soil and water conservation. In-row subsoiling is also used under controlled traffic conditions to minimize the potential for recompaction when subsequent cultural operations are performed. (Controlled traffic farming involves confining wheel-track traffic to specified zones where soil compaction may occur and preventing traffic and avoiding compaction on the crop zone.)

### *Lister*

A lister is a plow with a double moldboard that moves soil to both sides of the furrow formed as it is drawn through the soil. A lister leaves soil in a series of alternate beds (ridges) and furrows, and crops may be seeded either on the beds or in the furrows, depending on the crop being grown, soil water conditions, and so on. Beds and furrows formed by listing provide for soil and water conservation when properly oriented for the prevailing conditions (on the contour for controlling erosion by water and conserving water; perpendicular to prevailing wind for controlling erosion by wind).

### *Minimum Tillage*

Minimum tillage involves using the minimum number of primary and secondary tillage operations to meet the requirements of a crop under the prevailing soil and climatic conditions. The number of operations required often is fewer that the number used in conventional tillage. In some cases, minimum tillage qualifies as a type of conservation tillage, providing for major soil and water conservation in many cases (*see* **TILLAGE, *Conservation Tillage***).

### *Moldboard Plow*

A moldboard plow is a tillage implement that shatters the soil with partial to complete inversion of the surface layer to the depth of plowing, burying some or all surface residues. Moldboard plowing usually leaves the soil surface relatively rough, thereby initially resulting in conditions favorable for controlling erosion by water and wind and conserving water. It is, however, a form of clean tillage (*see* **TILLAGE, *Clean Tillage/Clean Culture***) and may under some conditions result in severe soil or water losses when used in conjunction with secondary tillage.

### *Mulch Farming/Tillage*

*See* **TILLAGE, *Stubble Mulch Tillage.***

### *Mulch Treader*

*See* **TILLAGE, *Skew Treader.***

### *No-Tillage*

No-tillage is the practice of planting a crop without any tillage for seedbed preparation following harvest of the previous crop. No-tillage is also known as zero-tillage, no-til or no-till, slot planting, direct drilling, and direct seeding. Crop seeding with no-tillage is accomplished with appropriate planters or seeders capable of opening a slit in the soil to the desired depth, placing the seed, and satisfactorily closing the opening to provide good seed-soil contact for seed germination and seedling establishment. In some cases, subsoiling is carried out in conjunction with the planting operation to disrupt a dense soil layer and thus achieve improved plant rooting (*see* **TILLAGE, *In-Row Subsoiling***). No-tillage planting also has been accomplished by aerially seeding the next crop before harvesting the present crop. Such planting strongly depends on timely precipitation for successful seed germination and crop establishment. Some no-tillage planting has been carried out by punching a hole in the soil. For a no-tillage system, the crop usually is not cultivated and chemicals are used to control all weeds. No-tillage is a form of conservation tillage (*see* **TILLAGE, *Conservation Tillage***) and is not recommended when the minimum required crop residues (30 percent cover after crop planting) are not available. No-tillage sometimes allows more timely crop establishment because planting can take place without waiting for seedbed preparation operations. Because most crop residues are retained on the surface, the no-tillage method has proven highly effective for controlling erosion by water and wind. It also improves water conservation under some conditions. In humid regions on poorly drained soils, the poor drainage problem may be increased because of reduced runoff and evaporation when no-tillage is used.

### *Paraplow*

A paraplow is a subsoiling implement designed to achieve greater lateral soil shattering than is obtained with a straight-shank subsoiler. The greater shattering results from broad lifting surfaces mounted at the bottom of angled shanks that are drawn through the soil. The increased shattering improves conditions for water infiltration and conservation and improved root penetration under some conditions, thereby also decreasing the potential for erosion by water and wind.

### *Plowless Farming*

*See* **TILLAGE, *No-Tillage; Stubble Mulch Tillage.***

### *Reduced Tillage*

A reduced tillage system is one in which the total number of operations used to prepare a soil for a crop is reduced from that normally used under the prevailing soil and field conditions. In some cases, tillage and planting are accomplished in one pass through the field (*see* **TILLAGE, *Minimum Tillage***).

### *Ridge Tillage*

Ridge tillage is a system involving ridges being reformed on the planted row by cultivations during the growing season of a crop, with the next crop then being planted into the ridges formed during the previous growing season. With ridge tillage, most crop residues usually remain on the surface, which is beneficial for conserving soil and water.

### *Rod Weeder*

A rod weeder is an implement with a straight, square rod (or bar) that rotates longitudinally below the surface as it is drawn through the soil. The rod, powered by the tractor pulling the implement or by linkage to wheels on the implement, rotates backward relative to the direction of travel. This action pulls or cuts roots from weeds with minimum disturbance of the soil surface, thereby retaining most crop residues on the surface for soil and water conservation purposes.

### *Rotary Cultivator/Rotary Weeder*

*See* **TILLAGE, *Skew Treader.***

### *Rotary Hoe*

A rotary hoe may be a ground-powered implement such as a skew treader (*see* **TILLAGE, *Skew Treader***) or a tractor-powered implement such as a rotary tiller or rotovator (*see* **TILLAGE, *Rotary Tiller/Rotovator***). The implement is used for shallow tillage to control weeds and loosen, shatter, or mix the soil. Use of a rotary hoe, especially the tractor-powered type, leaves the surface materials pulverized, and thus highly susceptible to erosion by water and wind and unfavorable for water conservation under some conditions.

### *Rotary Spader*

A rotary spader is an implement with several "spades" mounted on a heavy drumlike shaft. As the implement is drawn through the field, the spades penetrate and turn the soil, leaving it in a relatively rough condition similar to that achieved by spading by hand. This usually provides for temporary water storage on the surface, thereby reducing runoff and erosion by water and improving conditions for water conservation. The rough surface also reduces the potential for erosion by wind.

### *Rotary Tiller/Rotovator*

A rotary tiller is an implement that loosens, shatters, or mixes soil using rotary-motion power derived from the tractor. Rotary tillage leaves the soil surface smooth. The surface consists of finely divided materials, which provides a good seedbed but leaves the soil highly susceptible to erosion by water and wind under some conditions and may be detrimental to water conservation.

### *Sand Fighter*

A sand fighter is a specialized implement used to control erosion by wind. As the implement, which has pointed teeth mounted on a long shaft, is pulled over a field, the soil crust is broken and clods are left on the surface. Sand fighters are wide and are operated at relatively high speeds, which allows large areas to be treated quickly in an emergency—for example, to control erosion by wind that is in progress (*see* **TILLAGE, *Emergency Tillage***).

### *Scarifier*

A scarifier is any implement with sharp teeth that loosens soil at the surface to a depth of ~5 to 8 cm by a raking action as the implement is drawn through the field. Scarifying results in increased surface roughness, which may reduce the potential for erosion by water and wind and possibly improve conditions for water conservation.

### *Skew Treader*

A skew treader is an implement with curved tines. It uses ground-driven motion to shatter, loosen, or mix the soil surface. When it is drawn in one direction, the tines penetrate soil to a shallow depth (~75 mm) to control weeds, which conserves water, or to roughen the surface (increase cloddiness), which helps control erosion by wind. When the skew treader is drawn in the other direction, the surface soil is firmed (clods are broken and large voids are eliminated) to improve conditions for seed germination and seedling establishment. A skew treader operated at high speed is used to spread crop residues evenly over the soil surface to help control erosion by wind. Evenly spread residues also help control erosion by water and improve water conservation.

### *Slit Tillage*

Slit tillage involves the use of narrow, straight coulters or knives to open narrow slits in a soil to a depth beneath a layer that restricts root penetration, thereby allowing roots of seedlings from precision-planted seeds to penetrate the restricting layer and develop in the soil beneath. Slit tillage avoids the use of energy-intensive tillage needed for large-scale disruption or shattering of a restricting layer in the profile, which could increase the potential for erosion and reduce the potential for water conservation.

### *Strip Tillage*

Strip tillage essentially consists of tilling parallel bands of soil with the intervening bands remaining largely undisturbed. Depending on the condition of the undisturbed soil (surface aggregation, presence of plants or crop residues, etc.), use of strip tillage may increase the potential for soil and water conservation.

### *Stubble Mulch Tillage*

Stubble mulch tillage is the practice of preparing land for a crop in such a way that plant residues or other materials remain on the surface to provide protection against erosion by water and wind and to improve water conservation during the interval between crops and at least partly into the growing period of the next crop. It is a form of subsurface tillage accomplished with implements, such as a blade or duck-foot cultivator, that do not invert the surface soil layer as would be the case with a disk or moldboard plow (*see* **TILLAGE, *Blade Plow; Duck-Foot Cultivator [Field Cultivator]; Subsurface Tillage; V-Sweep***). Stubble mulch tillage is synonymous with conservation tillage (*see* **TILLAGE, *Conservation Tillage***), provided 30 percent of the soil surface is covered with residues. Stubble mulch tillage is used to control erosion by wind in many places. Even when fewer surface residues are available, stubble mulch tillage may provide some protection against erosion by water and wind and some water conservation benefits because of the surface roughness and depressions that result from the use of stubble mulch tillage implements.

### *Stump-Jump Chisel Plow*

A stump-jump chisel plow is a specially designed chisel implement for loosening and/or roughening a soil without causing damage to the implement when it strikes tree stumps or boulders. Loosening the soil and roughening the surface reduces the potential for erosion by water and wind and may improve conditions for water conservation.

### *Subsoiling*

Subsoiling is the use of any implement that loosens or shatters a soil to a depth usually greater than 40 cm without inverting the layers and with minimum mixing of the soil. The aim of subsoiling often is to disrupt a dense layer in the profile, thereby improving conditions for water infiltration and root growth to greater depths. Greater water infiltration reduces runoff and the potential for erosion by water and improves conditions for water conservation. The potential for erosion by wind is reduced if subsoiling increases roughness of the soil surface.

### *Subsurface Tillage*

Subsurface tillage is any tillage that loosens or shatters a soil with minimum disturbance of the soil surface (*see* TILLAGE, ***Blade Plow; Stubble Mulch Tillage; Subsoiling***). Most crop residues usually are retained on the soil surface and the surface may be roughened, thereby providing conditions favorable for soil and water conservation.

### *Tied Ridges/Tied Ridging*

*See* TILLAGE, ***Furrow Diking.***

## *Timeliness*

Timely tillage is highly important for preparing fields for timely crop planting and establishment and to rapidly achieve adequate soil cover to protect against erosion by water and wind, improve conditions for water conservation, and provide conditions for optimum crop productivity. Timely tillage is important also for controlling weeds, thereby conserving soil water for use by crops. If erosion by wind is anticipated or in progress, timely emergency tillage can provide protection against erosion and even possibly avoid damage to small crop plants.

## *Trash Farming*

Trash farming is a term sometimes used to denote farming where crop residues are retained on the soil surface (*see* **TILLAGE, *Conservation Tillage; No-Tillage; Stubble Mulch Tillage*)** as opposed to the trash-free surface conditions achieved where clean tillage (*see* **TILLAGE, *Clean Tillage/Clean Culture***) is used.

## *Trench Plowing*

Trench plowing was a form of deep plowing used in early efforts to control erosion. It involved the use of two plows, with one plow opening a furrow and the other following in the same furrow at a greater depth.

## *V-Sweep*

A tillage implement equipped with a V-sweep is used to undercut the soil surface to prepare a seedbed and control weeds while retaining most crop residues on the surface or to cultivate an established crop. The sweeps have the general shape of a "V" and are attached to the implement either at the point or at the ends of the V. V-sweeps for seedbed preparation and weed control range from ~0.75 m to more than 2 m in width. Implements with V-sweeps are commonly used for tillage aimed at retaining crop residues on the surface to help control erosion by water and wind and to enhance water conservation (*see* **TILLAGE, *Stubble Mulch Tillage***). Implements for cultivating crops often have sweeps in the range 0.10 to 0.30 m in width.

### *Winged Subsoiler*

A winged subsoiler has subsoiling shanks modified by attaching wings to the shanks at a relatively short distance from the point. Use of a winged subsoiler generally results in greater loosening or shattering of the soil than occurs with a subsoiler without wings. The greater loosening may result in reduced runoff and erosion by water, greater water conservation, and improved soil conditions for root growth.

### *Zero Tillage*

*See* **TILLAGE, *No-Tillage.***

### *Zone Tillage*

*See* **TILLAGE, *Strip Tillage.***

## TRAFFIC RUTS

Vehicles (trucks and cars) or farm machinery (tractors, harvesters, etc.) traveling across rangeland, pastures, and fields under some conditions (e.g., when the soil is wet) often cause ruts that may provide a channel for runoff when subsequent precipitation occurs. Ruts in fields generally can be filled when tillage is performed, but special efforts may be needed to prevent gully formation in ruts on rangeland and in pastures.

# TRANSPIRATION

Transpiration is the normal process by which plants release water in a vapor form into the air. The water is taken up by plant roots from the soil. Transpiration is the plant contribution to evapotranspiration (*see* **EVAPOTRANSPIRATION**). If the volume of water transpired can be reduced, some water can be conserved for later use by the plants or by other plants grown later at the same location. Some reduction in transpiration is possible by applying antitranspirants to plants (*see* **WATER CONTROL PRACTICES, *Antitranspirant***). Transpiration can also be reduced by harvesting crops for their grain, fiber, and so on when they become physiologically mature and then preventing regrowth of the plants (*see* **MANAGEMENT, *Early Crop Harvesting***) or by growing crops (e.g., corn) for silage rather than for grain. Because noncrop plants (weeds) also transpire water, eliminating them from the crop production enterprise will reduce the volume of water transpired, thereby conserving water, which through subsequent improved plant growth may also provide erosion control benefits.

# U

## UNDERFLOW

Underflow is the flow of sediment-laden water into reservoirs, which results in siltation of the reservoirs. The underflow influences the site at which sediments carried in the stream are deposited in the reservoir.


© 2006 by The Haworth Press, Inc. All rights reserved.
doi:10.1300/5678_18

# WATER CONTROL PRACTICES

For the purposes of this book, water control practices are those that control or alter the flow rate or movement of water in streams, across the soil surface, from the soil (evaporation), or from plants (transpiration). The goal of such practices is to reduce the potential for erosion by water and to conserve water.

## *Antitranspirant*

An antitranspirant is a substance applied to plant leaves that reduces the amount of water transpired. It has the effect of reducing the amount of water extracted from soil by plants, thus conserving water for later use by these or subsequent plants grown on the same area.

## *Brick Weir*

A brick weir is a semipermanent weir placed in a gully to trap sediments and retain water long enough for vegetation to become established. Cheap bricks for such weirs can be made by mixing suitable sand present in the gully with cement. Other cheap bricks are those rejected for building purposes and those from nearby commercial brickyards.

## *Bunding*

Bunding involves constructing earthen dikes or establishing vegetative strips (bunds) to minimize or prevent water flow to adjacent land areas and to more uniformly retain water within the area surrounded by

© 2006 by The Haworth Press, Inc. All rights reserved.
doi:10.1300/5678_19

the bund. For more uniform water distribution, land within the bunded area often is leveled. Controlling water flow, besides conserving water, provides erosion control benefits.

*Broad-Base Bunds*

Broad-base bunds are large and usually require special equipment for their construction. Such bunds result in greater loss of land for crop production than narrow-base bunds.

*Contour Bunds*

Contour bunds are constructed across the slope of the land. When combined with contour tillage or land leveling, they provide for maximum water retention and, therefore, maximum water conservation. Contour bunds are most adaptable to semiarid or subhumid regions where water conservation is highly important for successful crop production.

*Graded Bunds*

Graded bunds are mainly suited to regions where excess water must be removed from the land. When graded bunds are combined with graded-furrow tillage, the water is removed at nonerosive flow rates. Slow water flow from land has been shown to increase water storage in soil and water levels in wells.

*Grass Bunds*

Grass bunds are narrow strips of grass used in configurations similar to earthen bunds, namely, contour or graded. The grass forms a dense barrier that retards water flow downslope. By trapping sediments, such bund tends to result in leveling of the land between adjacent bunds. Vetiver grass has been shown to be highly effective for controlling erosion when used as a bund (*see* **BARRIERS, *Vetiver Grass***).

*Narrow-Base Bunds*

Narrow-base bunds are small earthen dikes usually less than 0.7 m wide. They may be constructed by hand or with appropriate equipment at a relatively low cost. One advantage of narrow bunds is that the amount of land area devoted to bunds is small.

***Conduit***

A conduit is a pipe or channel used to convey water. Conveying water through a pipe eliminates erosion and water losses en route. In contrast, erosion may occur and some water loss usually occurs when water is conveyed in a channel. Water conveyed in a lined channel (e.g., with concrete) does not cause erosion, but some water normally is lost through evaporation. Use of an unlined channel may result in some erosion, and water may be lost by evaporation, seepage through channel walls, and through use by plants growing beside the channel (*see* **PHREATOPHYTE**).

***Crossties***

Crossties are raised mounds of soil between parallel ridges of a furrow-diking (or tied-ridging) system. In conjunction with the ridges, crossties result in rectangular depressions for retaining water uniformly over the entire area, thereby preventing runoff that could cause erosion by water and retaining the water for storage in soil (*see* **TILLAGE, *Furrow Diking***).

***Dams***

A dam is a barrier placed in a stream or site where water normally flows to store water, raise the water level for diversion, create a hydraulic head, control or prevent gully erosion, or retain soil, rock, or other debris.

### *Brush/Post Dam*

Brush/post dams are structures built in a gully to slow the flow of water downslope. A one-post brush/post dam consists of one row of posts placed securely into the soil at a spacing of ~1 m, with brush packed behind the posts (at the upstream side) and secured to the posts. For a two-post brush/post dam, two rows of posts ~1 m apart are used, with brush packed between the rows.

### *Brushwood Dam*

Brushwood dams are constructed by tightly packing brushy materials across the channels of gullies. The materials are placed between two rows of vertical stakes driven into the ground. They are then held in place by tying the stakes together with wires or by placing larger materials on top of the brushy materials. Brushwood dams are used mainly to control gully erosion and to trap silt being carried downstream in the gully.

### *Check Dam*

A check dam (or gully plug) is a structure used to control erosion in gullies while vegetation is being established to control the erosion on a permanent basis and to retain enough soil and water to provide favorable growth conditions for the protective vegetation. Materials used to construct temporary check dams include brush, poles, loose rock, and wire. Permanent check dams may be constructed of reinforced concrete, masonry, cemented rocks, earth materials (with proper construction), or metal.

### *Concrete/Masonry Dam*

Concrete or masonry dams provide for water retention and erosion control on a permanent basis, provided they are properly designed and constructed. Such dams may range in size from small check dams to control erosion in gullies to large dams on streams to help control floods and retain water for subsequent use for agricultural, municipal, industrial, power generation, and recreational purposes.

*Detention Dam*

The primary use of detention dams is to retain water, often for agricultural purposes such as water for livestock (farm ponds) and for irrigating crops. Large dams, however, also detain water for other purposes (*see* **WATER CONTROL PRACTICES, *Dams,*** *Concrete/Masonry Dam*).

*Earthen Dam*

Farm ponds and flood control structures usually are constructed of earthen materials. The dam site for such structures requires proper preparation to minimize the potential for seepage where the earthen fill material is placed on the existing earthen material. The fill material is excavated upstream from the dam being constructed, thereby increasing the capacity of the resultant reservoir. Earthen dams require a well-designed spillway to convey excess water from the reservoir during major flow events, thereby reducing the likelihood of the dam being destroyed by the water flowing from the reservoir.

*Flood Control Dam*

Flood control dams usually are series of dams located in the major drainage ways that flow into the stream flowing through a larger watershed. The dams retain water from major rainstorms, reducing the potential for flooding of the stream. Flood control dams are designed to slowly release most of the water initially retained, providing maximum retention capacity for a subsequent rainstorm. In addition to reducing the potential for flooding, releasing water slowly reduces the potential for erosion of streambanks and of land adjacent to the stream if the stream overflows its channel.

*Furrow Dam*

Furrow dams (also known as dikes or blocks) are small earthen dams placed in furrows of cropland. Their main purpose is to retain water from rain or sprinkler irrigations more uniformly over the entire field for direct use by the crop or for storage in soil for later use by a crop. Because the water is retained on the land, the potential for erosion is greatly reduced, except during major storms when the retention capacity of the dams is exceeded (*see* **TILLAGE, *Furrow Diking***).

### *Gully-Head Dam*

Gully-head dams are placed below the head of an actively eroding gully that is steadily eating its way upstream. Water ponded behind the dam submerges the head of the gully, which results in the energy of water being dissipated as it flows into the pond.

### *Log Dam*

Log dams are of two types. One type is constructed in a manner similar to that used for brushwood dams. The logs are packed between two rows of sturdy upright posts driven into the bed of a gully. The logs are then held in place by strong wires strung between the vertical posts (*see* **WATER CONTROL PRACTICES, *Dams,*** *Brushwood Dam*). A second type of log dam consists of a single row of posts set adjacent to each other to form a wall. The posts should be fixed deep in firm soil in the bed of the gully to resist being carried away by water. Cross-members attached to the posts can improve the rigidity of the structure. Both types of log dams should be constructed with a provision (central notch or low area) to allow water to flow over part of the dam under high-flow conditions.

### *Netting Dam*

Netting dams are types of low check dams usually placed near the starting point of a gully. Posts are driven into the ground at or near the sides of the gully, and wire netting is strung between the posts, with the lower edge of the wire buried in the soil. Straw or brush is loosely placed against the upstream side of the netting. The barrier formed by water packing those materials against the netting slows the flow of water, which traps sediments carried by the water, thereby reducing sediment transport to streams.

*Regulating Dam*

A regulating dam is a type of permanent flood control dam (*see* **WATER CONTROL PRACTICES, *Dams,*** *Flood Control Dam*). The reservoir behind the dam has sufficient capacity to retain water from a single storm, and the water drains through a relatively small, permanently open pipe extending through the dam in 1 or 2 days, depending on the amount of water retained. After drainage, the empty reservoir is ready to capture water from the next storm. Capturing the water and releasing it slowly prevents flooding of the main stream of the watershed, thereby reducing the potential for erosion of the main channel and adjacent lands.

*Sand Dam*

A sand dam is a weir, usually constructed of masonry across an ephemeral river or watercourse to trap sand and store water. Sand dams serve as a source of water during the dry season in semiarid regions. Approximately 30 percent of the sand volume may be filled with water when flow in the stream stops. Some of the water will be lost by evaporation, but less than is lost from an open reservoir. Evaporation largely ceases when the water level drops to approximately 50 cm from the surface. The remaining water provides for the growth of forage grasses planted beside the streams. The trapped sand raises the bed of the streams, thereby reducing erosion of the streambanks.

*Silt-Trap Dam*

A silt-trap dam (also known as a silt basin or sedimentation basin) is any dam that forms a basin or pond at the upstream end of a channel or reservoir where sediment-laden water is held for a sufficient period of time to allow sediments to settle. This reduces the amount of sediment carried downstream into streams or reservoirs.

*Slab Dam*

A slab dam is a low dam constructed from slabs (first cuttings of logs sawed for lumber) in a small gully. The slabs are held in place by poles anchored in the soil, with the outer ends of the slabs embedded in the banks of the gully. The center portion of such dams is lower than the side portion, allowing water to flow over the dam without scouring the soil at the sides.

#### *Woven-Wire Dam*

A woven-wire dam is constructed of wire netting held in place by poles anchored in the soil of a gully. Brush or straw placed on the upstream side of the wire reduces the rate of water flow in the gully, thereby reducing the potential for erosion (*see* **WATER CONTROL PRACTICES, *Dams,*** *Netting Dam*).

### ***Detention Reservoir***

*See* **WATER CONTROL PRACTICES, *Dams,*** *Detention Dam.*

### ***Dike***

A dike is an embankment or a levee used to confine or control water, such as one built beside a stream to avoid the flooding of adjacent lowlands. Dikes are also sometimes used to confine irrigation water to a given area.

### ***Ditch***

Ditches are an essential feature of landscapes where excess water may be present in fields at some time before or during the growing season of a crop. The ditches carry the excess water downslope to streams.

#### *Drainage Ditch*

Properly designed and constructed drainage ditches can remove excess water from fields at nonerosive flow rates (*see* **WATER CONTROL PRACTICES, *Terrace,*** *Diversion Terrace;* ***Waterway***).

*Hillside Ditch*

A hillside ditch is one dug across the slope (on the contour) of land that is too steep for bench terracing. Soil and water are trapped in the ditch. During heavy rains, excess water overflows into a suitable outlet, often a stone-lined channel. Hillside ditches are commonly dug by hand and, therefore, are used almost exclusively where hand labor is readily available and cheap.

*Natural Ditch*

Natural ditches may or may not result in water flowing at rates that result in erosion, depending on the slope of the land and the amount of land in the watershed. The amount of land influences the amount of water that flows in the ditch: usually, the larger the watershed, the greater the potential for erosion.

***Drainage***

Drainage is the natural or induced flow of excess water from the landscape. Erosion is minimized if the water flow rate is slow. Although removal of excess water is the goal of drainage, water conservation may be desirable under conditions where the water has accumulated in low-lying areas without fully refilling the soil profile with water (e.g., during a large, high-intensity rainstorm). Under such conditions, retarding the natural flow of water from the landscape, other than through natural ditches, could improve water conservation over the entire landscape.

*Natural System*

Natural drainage occurs downslope through ditches (*see* **WATER CONTROL PRACTICES, *Ditch,*** *Natural Ditch*) or across the landscape. It also occurs through the soil profile, provided no restricting layers (horizons or other impermeable zones) are present in or below the soil profile.

*Open Channels*

Open channels may be constructed to remove excess water from land. To minimize the potential for erosion, they should be constructed at a grade that results in nonerosive water flow downslope.

*Outlets*

Drainage outlets are the points where water from ditches, channels, or subsurface drains in a field empty into downslope positions such as waterways and streams. Because relatively large volumes of water may flow from the area being drained and be emptied on a relatively small area, the potential for erosion by water often is high at such points. Thus, well-designed and constructed waterways or other structures often are needed to control water flow downslope.

*Subsurface Drainage*

Subsurface drainage may be natural drainage through the soil profile or more rapid drainage from induced flow. In the latter case, the water still flows to some depth in the profile, at which point it then flows laterally through drains installed in the soil. Mole drains are unlined openings formed by pulling a bullet-shaped cylinder through the soil. For tile drains, pipes made of polyvinyl chloride (PVC), concrete, burned clay, or similar materials, with openings through which water readily enters, are placed underground at an appropriate grade to collect and transport excess water from the landscape.

***Drains***

Special drains sometimes are needed when more water becomes available than can readily infiltrate the soils or where water accumulates from an impermeable surface such as a road or highway.

*Lock and Spill Drain*

A lock and spill drain (or ditch) is a type of terrace constructed on land that is too steep for bench terracing. The drain (~0.5 m wide and deep) is dug at a slight grade across the slope of the land. Slightly raised walls (locks) are left in the bed of the drain to form smaller basins to trap water and improve infiltration. When larger storms occur, the water overtops the locks and spills toward the outlet, which may be a stone-lined or similar channel that provides protection against erosion.

*Miter Drain*

Miter drains are used beside roadways in conjunction with road drainage ways to carry runoff nonerosively from the road. They are constructed at ~45° away from the road at intervals that depend on the expected intensity of rain in the region. Their purpose is to divert storm runoff water from the roadside ditch before unmanageable volumes of water accumulate that could result in serious erosion beside the road.

*Stormwater Drain*

A stormwater drain is a ditch or channel that intercepts stormwater or floodwater from upslope positions, thereby protecting downslope arable land against erosion or excess water. The water is diverted to suitable downslope outlets through which it is carried into a stream or reservoir. A stormwater drain is also known as a diversion terrace, diversion ditch, storm water channel, or storm drain (*see* **WATER CONTROL PRACTICES, *Terrace,*** *Diversion Terrace*).

*Tile Drain*

*See* **WATER CONTROL PRACTICES, *Drainage,*** *Subsurface Drainage.*

***Drop Structure***

A drop structure is used to drop storm runoff water in a gully nonerosively from one level to a lower level and to dissipate the surplus energy of the falling water. The structure may have a vertical or an inclined drop and is constructed of concrete, bricks, masonry, or similar materials (*see* **STREAM EROSION CONTROL, *Box Inlet***).

### *Earthen Channel*

An earthen channel (*see* **WATER CONTROL PRACTICES, *Waterway***) is a structure specially designed to convey storm runoff water at a nonerosive rate to a suitable natural waterway or to convey irrigation water from a suitable source to the point where it is applied to the land. Proper design based on land slope, expected flow rate and amount, and so on is essential to minimize the potential for erosion in the channel.

### *Furrow*

A furrow is a shallow channel at the soil surface that results from using a plow or disk to prepare the land for crop production, including the conveyance and distribution of irrigation water within a field. Depending on their orientation relative to the slope of the land, furrows may positively or negatively affect erosion by water and water conservation. When they are oriented along the slope of the land, water flow rates may be excessive, increasing the potential for erosion and decreasing the potential to use irrigation water efficiently and to conserve water. Orienting furrows at a slight grade across the slope reduces the water flow rate, thereby reducing the potential for erosion and providing more time for infiltration, which increases the potential to use irrigation water more efficiently and to conserve water. Furrows oriented across the slope (contour furrows) are not suitable for applying irrigation water, but they control runoff and conserve water because water from precipitation is retained on the land, except when excessive precipitation overtops the furrow ridges. Furrow orientation perpendicular to the direction of prevailing winds is often used to help control erosion by wind.

### *Levee/Dike (Embankment)*

*See* **WATER CONTROL PRACTICES, *Dike.***

### *Microwatershed*

A microwatershed is a small change in soil surface elevation designed to cause water to flow to and concentrate at a given site. It results in little soil wetting of the contributing area and wetting to a greater depth at the receiving site, thereby potentially providing more water to plants at the receiving site. Microwatersheds are beneficial in achieving some production in water-deficient regions.

### *Open-Channel Flow*

Open channels are used to collect and/or divert streams of water such as runoff from terraces, drainage from fields, and irrigation water to fields. To protect channels from erosion, they should have adequate cross section and depth and an appropriate gradient to handle the expected volume of water. These factors are to some extent influenced by whether the channels are lined or unlined. More rapid flow is possible in lined channels; thus cross section and depth usually can be reduced. Using lined channels also reduces water losses from seepage and use by plants (phreatophytes), which often are major factors when unlined channels are used to convey irrigation water.

### *Pond*

A pond is an earthen structure for holding water for use by animals or for irrigating crops (also called a reservoir). A pond is formed by excavating the earth and using the excavated material to form a dam across the drainage way. Under some conditions, a pond (tank) can be built in a stream without damming it; the excavated material is placed on the sides of the pond, allowing water to fill the pond and excess water to continue flowing downstream. Such ponds retain water during dry, rainless periods. A pond constructed in the channel at the downslope end of a graded terrace and equipped with a suitable subsurface drainage outlet can temporarily capture and store water that could cause erosion as the water flows from the terrace at a relatively fast rate. The water subsequently drains slowly from the pond and flows downslope at a nonerosive rate.

### *Pond Seals*

Water retention in a farm pond made by excavating a pit or constructing an embankment is strongly influenced by the earthen material in the pit and that used for the embankment. Excellent water retention occurs in some cases; water seepage occurs in others. To improve retention where seepage is a problem, the pond can be lined with a material such as bentonite, which is a highly plastic clay that swells significantly when it becomes wet. Another approach to preventing seepage is to line the pond with a manufactured liner such as one made of rubber.

### *Porous Check*

A porous check is a temporary gully control structure designed to slow the flow of water and to cause silt deposition where vegetation is being established. It can be a rock bolster, brushwood dam, log dam, or brick weir (*see* **STREAM EROSION CONTROL,** ***Bolster [Rock];*** **WATER CONTROL PRACTICES,** ***Brick Weir; Dams,*** *Brushwood Dam, Log Dam*).

### *Runoff Collection/Recycling*

The goal of runoff collection and recycling is water conservation, but it has indirect effects on erosion by water and wind through the subsequent use of that water for crop establishment, growth, and production. Runoff can be collected and stored in ponds or reservoirs. It can then be pumped to irrigate a crop on the land from which it was collected, applied to a downslope crop either by pumping or through gravity flow, or used as a water supply for animals or nonagricultural purposes.

### *Snow Trapping*

The stubble of wheat and other crops can be managed to increase water conservation where snow constitutes a significant portion of the potential water supply for crop production. The characteristics of stubble that affect snow trapping include its overall height, alternating strips of tall and short stubble, and stem number (*see* **CLIMATE,** ***Rain/Other Precipitation;*** **RESIDUES,** ***Crop Residues,*** *Orientation*).

### *Spillway/Emergency Spillway*

A spillway is a natural or specially constructed open or closed channel (or both) used to convey excess water from a reservoir. It may have gates to control the flow rate of water from the reservoir or to stop the flow under some conditions. To minimize the potential for erosion, the channel should be constructed of nonerosive materials (e.g., concrete) or have grass established in it and be wide enough and at a grade that allows water flow at a nonerosive rate. An emergency spillway provides for the flow of excess water safely through or around a dam when the primary spillway fails to function properly or when more water must be removed than can be removed through the primary spillway. It should have design features similar to those of the primary spillway to minimize the potential for erosion. An apron, which is a floor or lining, may be installed to protect the surface from erosion where water flows from a spillway.

### *Straw Wattle (Fiber Roll)*

A straw wattle (or fiber roll) is constructed of such materials as rice straw. It is used to trap sediments, protect storm drain inlets, direct runoff water to detention reservoirs, reduce flow velocities in channels, and control sediments at the perimeter of, for example, construction sites. In addition to trapping sediments that could be deposited downslope or downstream in waterways or reservoirs, straw wattles also provide for water conservation when water is diverted to appropriate reservoirs (*see* **WATER CONTROL PRACTICES, *Wattle Fence***).

### *Syrup-Pan System*

A syrup-pan system is a water-spreading system used where dryland farming is practiced and where water from an outside source is available. It is most suitable where the surface slope is slight to moderate and where the soils are relatively deep, fertile, and permeable. The water is runoff from adjacent land (watershed) or, for example, from a road or highway. The system consists of a series of level terraces, with successive alternating terraces downslope having one end blocked and the opposite end not blocked. Water from the outside source is allowed to flow onto the land above the upper terrace. After wetting that area, the water flows through the open end of the first terrace, wets the next area as it flow to the opposite end, then flows around the second terrace, and so on. Because the water flows back and forth across the field as it moves downslope, the entire field can receive water other than through direct precipitation, providing additional water to the crop. If any water remains after wetting the lowest area, it should be discharged into a stabilized outlet.

### *Tailwater Control System (Buried Pipe)*

Tailwater is the runoff that occurs at the lower end of irrigated fields. Depending on the land slope at such points, runoff can cause severe erosion as it flows downslope. The potential for erosion is greatly reduced when the water is conducted downslope through buried pipes with appropriate inlets for water and an appropriate erosion control structure where the water flows from the pipe.

### *Tailwater Recovery System*

In irrigated cropping, tailwater is water that does not infiltrate the soil and flows from the lower end of the field. A tailwater recovery system collects that water and reuses it by pumping it to the initial site or an alternative site. This water could also be allowed to flow freely to irrigate a downslope area (*see* **WATER CONTROL PRACTICES, *Runoff Collection/Recycling; Tailwater Control System [Buried Pipe]***).

### *Tank*

*See* **WATER CONTROL PRACTICES, *Pond.***

### *Terrace*

From an agricultural viewpoint, "terrace" has several meanings. First, a terrace is a natural, relatively level, steplike surface beside a stream or shoreline that represents the former position of a floodplain, lake, or seashore. Second, a terrace is a raised and generally horizontal strip of earth or rock constructed on or nearly on the contour of a hill to make the land suitable for crop production and to prevent accelerated erosion. Third, a terrace is an embankment or a combination of an embankment and a channel constructed across the slope of the land to control erosion by diverting or temporarily retaining surface runoff water rather than permitting it to flow uninterrupted down the slope. This discussion deals with the second and third types, which are constructed with appropriate tools or machinery by humans. Terraces are classified according to their alignment, gradient, outlet, and cross section. Their purpose is to control erosion by water and/or conserve water by diverting or interrupting the flow downslope or holding water on land so it will infiltrate the soil.

*Absorption Terrace*

An absorption terrace is constructed on the true contour or flat (level) along its entire length and is designed to retain water so that it infiltrates the soil.

*Bench Terrace*

Several types of bench terrace are recognized. Some types are used on steeply sloping land (greater than about 15 percent), whereas others are used on relatively flat land (less than about 15 percent):

1. A back-slope terrace is constructed on steeply sloping land, and the channel (or the floor) of the terrace is lower adjacent to the hill than at its outer edge. As a result, water remains on the land unless the channel slope is graded toward an outlet or large amounts of precipitation are received. The wall between adjacent terraces is vertical or almost vertical.
2. Conservation bench terraces (also known as Zingg bench terraces) are used on land with a relatively gentle slope. The lower-slope area between adjacent terraces is leveled, with the remaining watershed area being left unleveled. Ratios of watersheds to leveled areas range from 1:1 to 6:1, with the appropriate ratio depending on expected precipitation and soil conditions. When runoff-producing precipitation occurs, runoff from the watershed is captured on the

leveled area, providing additional water that permits more intensive cropping on the leveled area. For example, with a 2:1 ratio and 40 mm of precipitation that results in 15 mm of runoff from the watershed, the leveled area would retain the 40 mm of precipitation and receive an additional 30 mm of runoff, resulting in 70 mm of water on the leveled area and only 25 mm on the watershed area. In the semiarid southern Great Plains of the United States, annual cropping has been possible on the leveled area where the ratio is 2:1. On the watershed area, a cropping system involving a fallow period between crops is used to store additional water for the next crop. Capturing the runoff also helps control erosion from fields where such terraces are used, but erosion by water is possible on the watershed area within the field.

3. Irrigation bench terraces are similar to level bench terraces except that they have a raised ledge at the outer edge to retain the irrigation water.

4. For a level bench terrace used on steeply sloping land, the channel (or floor) of the terrace is level from its outer edge to its back and the wall between adjacent terraces is vertical or almost vertical.

5. A mini-bench terrace is used on relatively flat land with the bench formed between narrow ridges (berms) of soil that hold the water on the land. Such terraces are primarily appropriate to dryland cropping conditions, as in semiarid regions, and usually are constructed on the contour of the land. The area between adjacent terraces is leveled in some cases. The space between adjacent terraces is equal to one or more widths of the implements used for crop production, allowing efficient operation of crop production machines and implements in the field.

6. Mini-conservation bench terraces are similar to conservation bench terraces, except that the distance between adjacent terraces is much narrower (e.g., three times the width of implements used), which reduces the construction costs associated with leveling a portion of the land.

7. Outward-sloping terraces are used on steeply sloping land. In contrast to a back-slope terrace, the channel (or floor) of an outward-sloping terrace is lower at its edge than adjacent to the wall. As a result, water can flow from an outward-sloping terrace. The wall between adjacent terraces is vertical or almost vertical.

8. A Reddick bench terrace is used for growing trees (e.g., citrus) along irrigation grade lines. Along these lines, a small ridge is made with two or three furrows on the upslope side. Trees are planted on the ridge and irrigations are applied in the furrows. Subsequent cultivations move soil toward the ridge but leave a strip of vegetation in the row of trees. By avoiding cross cultivation, a bench terrace is gradually formed across the hillside.

9. Step terraces are relatively narrow terraces on steeply sloping land, and their design is similar to that of level bench terraces. Because the benches are relatively narrow, they frequently are used for fruit, coffee, and tea trees or bushes for which regular cultivation is not required.

10. Zingg bench terraces (see Entry 2 above).

*Broad-Base Terrace*

Broad-base (or wide-base) terraces are wide enough that normal crop production operations used on areas between terraces can also be used on the terraces.

*Cattle Terrace (Trail)*

A cattle terrace is a pathway formed by cattle as they cross steeply sloping grazing lands. The pathways result from cattle repeatedly crossing the slope in their quest for forage. (It is easier for animals to walk across rather than up and down steep slopes.) Such pathways sometimes become the starting points for gullies, but little can be done to change the direction in which animals travel while grazing on such lands. Similarly, trails formed by animals repeatedly walking to or from a given point in pastures (e.g., their water supply or resting area) frequently become the starting point for gullies. In this case, when gullies begin to develop, some type of impediment can be placed in the trail so the cattle will walk in a different place and formation of a major gully is avoided. In extreme cases, barriers (e.g., fences) could be placed to force animals to traverse an area highly susceptible to erosion in a zigzag manner.

### *Channel Terrace*

A channel terrace is any terrace constructed by excavating a channel and using the excavated material to form an embankment on the downslope side of the channel.

### *Diversion Terrace*

A diversion terrace (diversion ditch, diversion drain, storm drain, storm water channel) consists of an individually designed channel with an embankment on the downslope side constructed across a hillside to protect downslope land from runoff from the hill or constructed above a terraced area to protect it from runoff from the unterraced area. Diversion terraces are also used to divert water out of active gullies, to protect farm buildings, to reduce the number of waterways, or with strip cropping to shorten the slope length so that the strips can more effectively control erosion.

### *Fanya Juu Terrace*

A fanya juu terrace is an intermittent terrace, developed or used in Kenya, that uses excavated soil to build an embankment upslope from the ditch with the goal of trapping silt to build a more level terrace. Silt trapping is enhanced by planting grasses or other vegetation on the edge or ridge of the terraces.

### *Fish-Scale Terrace*

A fish-scale terrace is a small terrace designed for an individual plant or a small number of plants. Such terraces are not connected to each other and are used on relatively steep land to capture water and trap sediments carried by the water.

### *Graded-Channel Terrace*

A graded-channel terrace is any terrace with a relatively small grade used to transport water downslope at a relatively slow rate. The grade usually is uniform for the entire length of the terrace, but variable-grade terraces are used under some conditions.

*Gradoni System*

The gradoni system is a type of terrace system developed to minimize reforestation problems. It has broad steps ~1.3 m wide and spaced ~7 m apart. The steps are constructed on the contour and are inclined inward at an ~30 percent slope to minimize erosion on the outer rim.

*Intermittent Terrace*

Intermittent (or orchard) terraces are small level or back-slope terraces spaced some distance apart and designed for individual rows of trees. An important requirement of such terrace systems on erosion-prone land is that a vigorous cover crop be grown on the area between the terraces.

*Level Terrace*

A level terrace may be a level bench terrace or a channel terrace that is level from end to end. Water retention on land is one major reason for using a level-channel terrace. Level-channel terraces may have one or both ends open to allow water to drain from behind the terrace, or ends may be closed (blocked) to retain potential runoff water on land for water conservation purposes.

*Litter Terrace*

A litter terrace develops on the land surface when litter carried by water is trapped or deposited behind some obstruction, thereby reducing the rate of water flow across the surface, providing more time for infiltration, and trapping sediments carried in water. (*Note:* Litter terraces are formed without human inputs.)

*Mangum Terrace*

A Mangum terrace is a terrace built up from both sides by repeated rounds with a suitable implement such as any plow that moves soil in one direction.

*Narrow-Base Terrace*

A narrow-base terrace is a terrace with a narrow ridge with steep sides. Such terraces cannot be crossed by crop production equipment and are too narrow to have a crop planted on them.

*Nichols Terrace*

*See* **WATER CONTROL PRACTICES, *Terrace,*** *Channel Terrace.*

*Orchard Terrace*

*See* **WATER CONTROL PRACTICES, *Terrace,*** *Intermittent Terrace.*

*Parallel Channel Terrace*

Parallel channel terraces are constructed so that the distance between adjacent terraces is uniform throughout their length. To construct such terraces, the channel gradient is allowed to deviate from the design gradient at some places. Constructing such terraces may also entail excavating soil to greater depths at high points and lesser depths at low points, but forming uniform embankments on the downslope side along the entire terrace.

*Platform Terrace*

Platform terraces are a type of intermittent terrace. They differ from the typical intermittent type, which is designed for rows of trees, in that they are small sections of terraces designed for individual plants. On erosion-prone land, a vigorous cover crop should be grown on the area between the terraces to reduce the potential for erosion.

*Rice*

*See* **WATER CONTROL PRACTICES, *Terrace,*** *Bench Terrace* (3).

*Ridge-type*

*See* **WATER CONTROL PRACTICES, *Terrace,*** *Narrow-Base Terrace.*

### *Terrace Outlet/Outlet Channel*

A terrace outlet is the place where water from above the terrace is discharged into a channel for conveyance from the field. Outlet channels usually receive water from one or more terraces and have a vegetative cover to minimize the potential for erosion due to the concentrated flow of water. In some cases, the channels may be tiled or have drop structures to reduce the potential for erosion.

### *Wadi Siltation*

Siltation that occurs in wadis may have negative or positive effects on soil and water conservation. The effect is negative when siltation reduces water flow in the wadi, thereby resulting in increased flooding that may lead to increased erosion of surrounding areas by water. In contrast, intentional silting achieved by constructing dams in wadis leads to water being stored in the sediments, which can be used by plants growing in the sediments. Such intentional silting is also used to spread water on surrounding areas and thus wet an area larger than that of the wadi itself. Such water spreading is viewed as a water conservation practice because it retains water on some of the land rather than allowing it to flow largely unimpeded downstream (*see* **WATER CONTROL PRACTICES, *Water Spreading***).

### *Water Harvesting*

Water harvesting is the practice of capturing runoff water from a given area and diverting it to a downslope area for immediate use or retaining it in a suitable reservoir for later application to another area. To enhance runoff, the contributing area may be smoothed, graded, compacted, treated to enhance crusting, or covered with plastic sheeting. Water harvesting generally is practiced in rainfall-deficient areas and the contributing area usually is much larger than the receiving area. In addition, the contributing area may not be suitable for plant growth (*see* **WATER CONTROL PRACTICES, *Syrup-Pan System; Terrace,*** *Bench Terrace* [2]; ***Water Spreading***).

### *Water Spreading*

Water spreading is the practice of diverting runoff water from gullies, channels, and waterways by means of dams, dikes, or ditches and spreading it onto relatively level downslope areas to promote plant growth. The diverted runoff water is retained on land rather than flowing uninterrupted downstream. Retaining water on land helps prevent soil movement into streams and reduces the amount of water entering streams, thereby reducing the potential for floods (*see* **WATER CONTROL PRACTICES, *Syrup-Pan System; Water Harvesting***).

### *Waterway*

A waterway is a natural or a specially designed and constructed channel for conveying runoff nonerosively downslope to a suitable outlet into a stream. A waterway may serve to convey water from terraces or to divert water away from buildings, farmsteads, and so on. To reduce the potential for erosion, waterways usually are planted to grass. In some cases, they may be lined with concrete or other nonerosive materials, especially on steeply sloping or highly erosive land. In special cases, suitable drop structures may be used to safely convey the water downslope.

### *Wattle Fence*

Wattle fences are constructed of materials such as straw, tree branches, twigs, logs, and boards. They serve to trap sediments or reduce offsite sedimentation by directing runoff to detention reservoirs, reducing water flow velocities where concentrated flow occurs, protecting storm drain inlets, and serving as an alternative to silt fences for sediment control at the perimeter of, for example, construction sites. Trapping sediments means that downslope/downstream siltation is reduced. When water flow is directed to reservoirs, water conservation is achieved.

### *Weir/Weep*

A weir (or weep) is a carefully constructed opening in a dike to allow water diverted from a gully, channel, or ditch to flow into a field when water spreading is being carried out. For relatively uniform water distribution, several weirs (weeps) generally are used for a given field. Further water spreading in the field can be achieved by strategically located furrows, ridges, and so on below the weirs.

### *Widely Spaced Furrows*

Furrow spacing is from ~0.75 to 1.0 m in many conditions where furrow irrigation of crops is practiced. For soils where lateral water movement from the irrigation furrow is adequate, the use of more widely spaced furrows (1.5 to 2.0 m) results in reduced wetting of the soil surface, leading to water conservation through reduced evaporative losses of applied irrigation water.

## WATERSHED

A watershed is the land within the area surrounded by natural ridges or hills or human-constructed surface elevations that result in all water within that area flowing to the same stream or low point.

Characteristics of a watershed such as size, surface slope, soil texture, and vegetative cover and its management strongly influence the potential for erosion by water and wind and for water conservation.

## WATER SHORTAGE

Water shortages are common in arid and semiarid regions, where they often result in limited plant growth unless irrigation water is applied. To achieve satisfactory plant growth without irrigation, water conservation practices are highly important. With satisfactory plant growth and careful management, the potential for erosion by water and wind can be reduced in semiarid and arid regions. Water shortages often occur also in some subhumid regions and less often in humid regions. When severe water shortages in these regions occur at critical plant growth stages, plant growth may be reduced so that the potential for erosion may be increased. Water conservation, therefore, usually is important in subhumid regions and sometimes even in humid regions.

## WATER TABLE FLUCTUATION

A water table is the zone beneath the soil surface that is saturated with water. The depth at which it occurs may strongly influence the potential for erosion by water and for water conservation. When the water table is near the surface, little potential exists for additional water to enter the soil, so runoff and the potential for erosion by water are increased. When the water table is at a greater depth, the potential for water infiltration and storage in soil is greater, which reduces the potential for runoff and erosion.

# WOODLAND

Woodlands are any lands used primarily for growing trees or shrubs. In addition to lands called forests, woodlands include shelter belts, windbreaks, wide hedgerows that contain woody plant species that provide food and shelter for wildlife, and tree-covered streambanks. Depending on tree and/or shrub type, density, and cover provided, erosion by water may or may not be a problem on woodlands. These factors also may affect water conservation. Erosion by wind usually is not a problem on woodland areas, and woodlands (shelter belts and windbreaks) consisting of woody species are used in some places to help control erosion by wind. Woodlands can be established through afforestation (planting trees on land previously not used for trees) or reestablished through reforestation. With proper management, woodlands can provide income to the producer or landowner through controlled grazing by livestock and wild animals, by harvesting products of the trees (e.g., nuts, fruits, resins, and juices), and by harvesting trees, bushes, and shrubs for poles, firewood, and timber.

Appendix

# Common and Scientific Names of Plants, Animals, and Other Organisms

The following list of common and scientific names is included to help readers identify plants, animals, and other organisms mentioned in the text that they may know by another name.

| **Common name** | **Scientific name** |
|---|---|
| Acacia | *Acacia* sp. |
| Alfalfa | *Medicago sativa* L. |
| Army worm | *Pseudaletia* sp. |
| Barley | *Hordeum vulgare* L |
| Bean | *Phaseolus* sp. |
| Bermuda grass | *Cynodon dactylon* (L.) Pers. |
| Cattle | *Bos* sp. |
| Citrus | *Citrus* sp. |
| Clover | *Trifolium* sp. |
| Coffee | *Coffea* sp. |
| Corn | *Zea mays* L. |
| Cotton | *Gossypium hirsutum* L. |
| Cricket | Various species |
| Deer | Various species |
| Earthworm | *Lumbricus* sp. |
| Goat | *Capra* sp. |

© 2006 by The Haworth Press, Inc. All rights reserved.
doi:10.1300/5678_20

| Common name | Scientific name |
|---|---|
| Grain sorghum | *Sorghum bicolor* (L.) Moench |
| Grasshopper | *Melanoplus* sp. |
| Hog | Various species |
| Jute | *Corchorus* sp. |
| Kudzu | *Pueraria thunbergiana* |
| Lespedeza | *Lespedeza striata* |
| Mite | Various species |
| Nematode | *Trichocephalus* sp. |
| Oat | *Avena sativa* L. |
| Pea | *Pisum* sp. |
| Peanut | *Arachis* sp. |
| Pearl millet | *Pennisetum glaucum* |
| Pineapple | *Ananas comosus* |
| Poplar | *Populus* sp. |
| Raccoon | *Procyon* sp. |
| Rice | *Oryza sativa* L. |
| Sheep | *Ovis aries* |
| Spider | Various species |
| Sugarcane | *Saccharum sp.* |
| Sunflower | *Helianthus annuus* L. |
| Tea | *Thea sinensis* |
| Termite | Various species |
| Vetch | *Vicia* sp. |
| Vetiver grass | *Vetiveria zizanioides* |
| Willow | *Salix* sp. |
| Winter wheat | *Triticum aestivum* L. |
| Wolf (wolves) | *Canis* sp. |

# Index

*Note to the reader:* The terms *conservation, water conservation, water erosion,* and *wind erosion* appear repeatedly in the text and are therefore not indexed.



© 2006 by The Haworth Press, Inc. All rights reserved.
doi:10.1300/5678_21

**Order a copy of this book with this form or online at:**
**http://www.haworthpress.com/store/product.asp?sku=5678**

**SOIL AND WATER CONSERVATION HANDBOOK**
**Policies, Practices, Conditions, and Terms**

_______in hardbound at $39.95 (ISBN-13: 978-1-56022-329-0; ISBN-10: 1-56022-329-4)

_______in softbound at $29.95 (ISBN-13: 978-1-56022-330-6; ISBN-10: 1-56022-330-8)

*232 pages plus index*

Or order online and use special offer code HEC25 in the shopping cart.

COST OF BOOKS_______

POSTAGE & HANDLING_______
*(US: $4.00 for first book & $1.50 for each additional book)*
*(Outside US: $5.00 for first book & $2.00 for each additional book)*

SUBTOTAL_______

IN CANADA: ADD 6% GST_______

STATE TAX_______
*(NJ, NY, OH, MN, CA, IL, IN, PA, & SD* residents, *add appropriate local sales tax)*

**FINAL TOTAL**_______
*(If paying in Canadian funds, convert using the current exchange rate, UNESCO coupons welcome)*

☐ **BILL ME LATER:** (Bill-me option is good on US/Canada/Mexico orders only; not good to jobbers, wholesalers, or subscription agencies.)

☐ Check here if billing address is different from shipping address and attach purchase order and billing address information.

Signature_______

☐ **PAYMENT ENCLOSED:** $_______

☐ **PLEASE CHARGE TO MY CREDIT CARD.**

☐ Visa ☐ MasterCard ☐ AmEx ☐ Discover ☐ Diner's Club ☐ Eurocard ☐ JCB

Account # _______

Exp. Date_______

Signature_______

Prices in US dollars and subject to change without notice.

NAME_______

INSTITUTION_______

ADDRESS_______

CITY_______

STATE/ZIP_______

COUNTRY_______ COUNTY (NY residents only)_______

TEL_______ FAX_______

E-MAIL_______

May we use your e-mail address for confirmations and other types of information? ☐ Yes ☐ No
We appreciate receiving your e-mail address and fax number. Haworth would like to e-mail or fax special discount offers to you, as a preferred customer. **We will never share, rent, or exchange your e-mail address or fax number.** We regard such actions as an invasion of your privacy.

*Order From Your Local Bookstore or Directly From*

**The Haworth Press, Inc.**

10 Alice Street, Binghamton, New York 13904-1580 • USA
TELEPHONE: 1-800-HAWORTH (1-800-429-6784) / Outside US/Canada: (607) 722-5857
FAX: 1-800-895-0582 / Outside US/Canada: (607) 771-0012
E-mail to: orders@haworthpress.com

**For orders outside US and Canada,** you may wish to order through your local sales representative, distributor, or bookseller.
For information, see http://haworthpress.com/distributors

*(Discounts are available for individual orders in US and Canada only, not booksellers/distributors.)*

PLEASE PHOTOCOPY THIS FORM FOR YOUR PERSONAL USE.

http://www.HaworthPress.com

BOF06

St. Louis Community College
at Meramec
LIBRARY